JULES STEWART is a journalist, historian and author. His books include *Madrid: The History; Albert: A Life; The Kaiser's Mission to Kabul* and *Gotham Rising: New York in the 1930s* (all published by I.B.Tauris); *Crimson Snow: Britain's First Disaster in Afghanistan*; *The Savage Border: The Story of the North-West Frontier*; *Spying for the Raj: The Pundits and the Mapping of the Himalaya* and *The Khyber Rifles: From the British Raj to Al Qaeda*. He lives in London.

To Ruby and Theo

'Jules Stewart's book is a graphic reminder of the errors that cast us so dearly in Afghanistan…and which must not be repeated today'
— **General Sir David Richards**

'…a compelling narrative history of Britain's three wars with Afghanistan…Stewart is an engaging wordsmith and writes of the most vivid episodes of Britain's wars of the nineteenth and early twentieth centuries with a fine style…'
— *International Affairs*

'Stewart manages to take a subject which can be fiendish in its complexity, and tell the story with great vividness and clarity, but also with scholarly authority. He has a fine eye for character and the revealing detail…a fine and readable introduction to the history of this field.'
— *Asian Affairs*

'extremely readable…Stewart has a journalist's talent for emphasising the main threads of the story'
— **British Commission for Military History newsletter**

'…successfully charts the ups and down of our relationship with that most difficult of countries…Stewart offers the reader a clear chronological review of the sorry saga…the author knows the country well and has written widely about it…'
— **The Guards Magazine**

'…the author paints an accurate picture of the political and military backgrounds of each incursion and the consequences …'
— **The Bulletin of the Military Historical Society**

'…a very interesting read that highlights how some lessons should never be forgotten.'
— *Soldier Magazine*

ON
AFGHANISTAN'S
PLAINS

The Story of Britain's Afghan Wars

Jules Stewart

New paperback edition published in 2019 by
I.B.Tauris & Co. Ltd
London • New York
www.ibtauris.com

First published in hardback in 2011 by I.B.Tauris & Co. Ltd

Copyright © 2011 Jules Stewart
Copyright Foreword © 2019 General Sir David Richards

The right of Jules Stewart to be identified as the author of this work has been asserted by the author in accordance with the Copyright, Designs and Patents Act 1988.

Image credit:
The Last Stand of the 44th Regiment at Gundamuck, 1842.
Painted by William Barnes Wollen.
Image courtesy of the Essex Regiment Museum, Chelmsford.

All rights reserved. Except for brief quotations in a review, this book, or any part thereof, may not be reproduced, stored in or introduced into a retrieval system, or transmitted, in any form or by any means, electronic, mechanical, photocopying, recording or otherwise, without the prior written permission of the publisher.

Every attempt has been made to gain permission for the use of the images in this book. Any omissions will be rectified in future editions.

References to websites were correct at the time of writing.

ISBN: 978 1 78831 416 9
eISBN: 978 0 85773 027 5

A full CIP record for this book is available from the British Library
A full CIP record is available from the Library of Congress

Library of Congress Catalog Card Number: available

Printed and bound by CPI Group (UK) Ltd, Croydon, CR0 4YY

Contents

List of illustrations	vi
Foreword by General Sir David Richards	xiii
Acknowledgements	xvi
Introduction	xvii
1 Cradle of Political Insanity	1
2 Victoria's First War	29
3 The Present Happy Moment	57
4 Vengeance is Mine, Sayeth Lord Ellenborough	89
5 The Pure Instinct of Dominion	115
6 Chronic Suspicion and Undignified Alarm	131
7 Nothing but Misfortune and Disaster	163
8 Once More unto the Breach	199
Notes	233
Bibliography	247
Index	251

Illustrations

Maps
1   Afghanistan and North-West Frontier   ix
2   Advance of the Army of the Indus into Afghanistan, 1838   x
3   British Retreat from Kabul to Jalalabad, 1842   xi

Plate section
1   Ahmad Shah Durrani, Founder of the Afghan Empire in 1747
2   Dost Mohammed with his youngest son. One of Afghanistan's wisest emirs, Dost Mohammed ruled from 1826 to 1863. He was deposed in 1839 by the British, who placed Shah Shuja on the throne, and was restored to power in 1842
3   Shah Shuja
4   Lady Florentia Sale, whose diaries revealed the tragic events that led to the Army's destruction on the retreat from Kabul
5   Alexander Burnes, the Government's envoy to the Court of Dost Mohammed, was the first high-profile victim of the Kabul uprising
6   Akbar Khan was Dost Mohammed's favourite son, and was responsible for the Army's massacre
7   William Macnaghten, the Government's chief political representative in Kabul, was murdered by Akbar Khan
8   *The Last Stand of the 44th Regiment at Gundamuck, 1842.* Painted by William Barnes Wollen. Image courtesy of the Essex Regiment Museum, Chelmsford
9   A young Pashtun tribesman awaits his victims in an Afghan Pass
10 Sher Ali Khan, known as the Iron Emir, was Afghanistan's longest ruling king, who occupied the throne from 1863 to 1879. He died in exile after the British launched their second invasion of Afghanistan

11 Field Marshal Lord Roberts of Kandahar, the legendary military figure who led the famous march from Kabul to Kandahar, where he defeated the Afghans and won the Second Afghan War
12 The Battle of Maiwand turned into a disastrous rout that cost the British more than one thousand casualties. It was one of the few instances in the 19th century of Asian forces achieving victory over a European Army. Image courtesy of the National Army Museum
13 A Russian stands in frustration at the door to British India, padlocked out by the Treaty of Gandamak
14 After the massacre of the Cavagnari Mission in Kabul, the Afghan is fed a portion of British rule
15 Badges in the Khyber Pass of British and Indian army regiments that fought in the Afghan wars and on the North-West Frontier
16 A caravan negotiating a narrow pass in Afghanistan. These caravans were also used to transport smuggled weapons to Pashtun fighters
17 King Amanullah invaded British India in 1919, touching off the Third Afghan War. The Emir lost the war but claimed victory for having recovered Afghanistan's full independence. He was forced to flee his country in 1929
18 The Khyber Pass, the major trade and invasion route between Central Asia and India
19 A street in the Peshawar bazaar around the time of the Third Afghan War
20 Royal Air Force BE2C biplanes, used to bomb Afghan positions in the Third Afghan War and later against rebellious Pashtun tribesmen on the North-West Frontier
21 A newspaper illustration of the RAF bombing Kabul's Bala Hissar fortress in the Third Afghan War
22 The border crossing between the North-West Frontier and Afghanistan. Travellers are warned to leave here on the return journey and be back in Jamrud no later than 5:00pm. Image courtesy of the Field Family Collection

When you're wounded and left on Afghanistan's plains,
And the women come out to cut up what remains,
Jest roll to your rifle and blow out your brains
An' go to your Gawd like a soldier.

From 'The Young British Soldier' by Rudyard Kipling

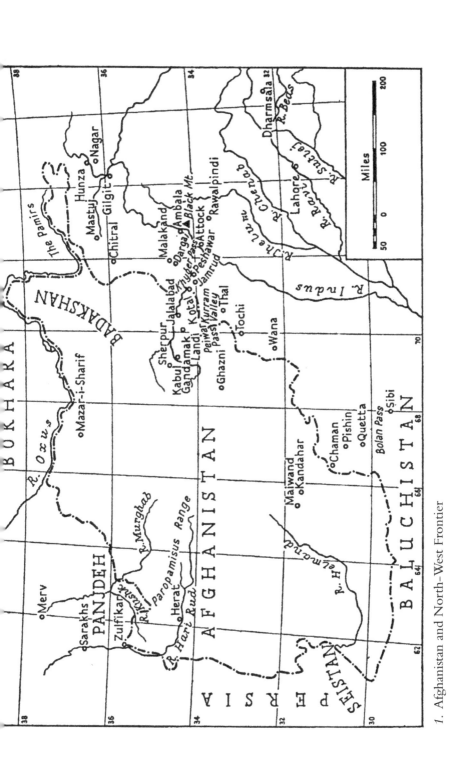

1. Afghanistan and North-West Frontier

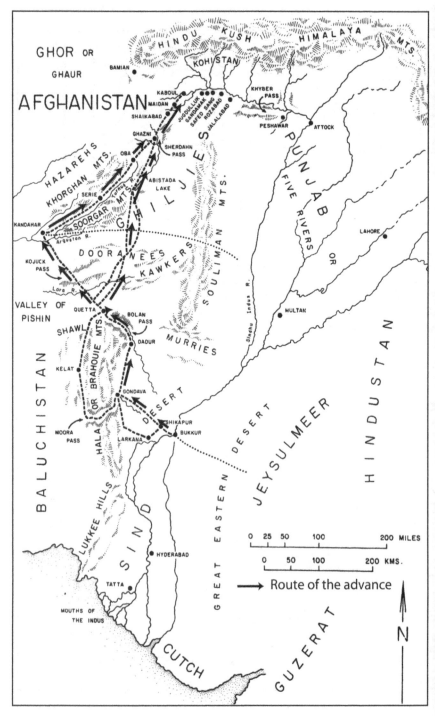

2. Advance of the Army of the Indus into Afghanistan

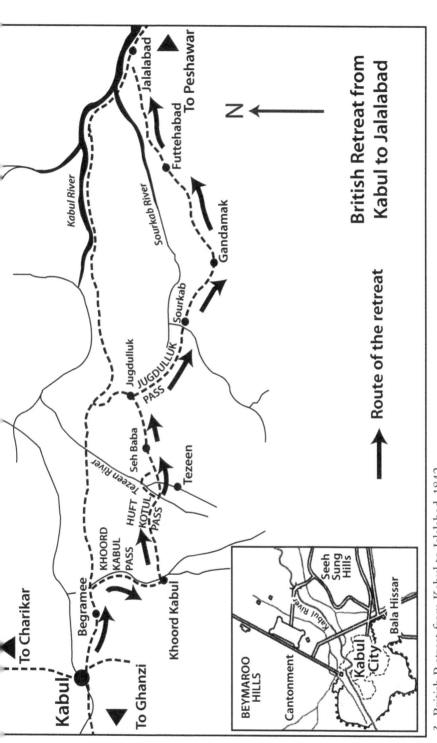

3. British Retreat from Kabul to Jalalabad, 1842

Foreword

General Sir David Richards

Afghanistan's military and political situation has entered a critical phase as President Ashraf Ghani has now recognised the need to engage with the Taliban to negotiate a sustainable peace. This is not an admission of weakness or defeatism on the part of the Afghan government. On the contrary, those militants who can be persuaded to sit down and talk to the civilian and military authorities are in fact acknowledging a reality, namely that victory is not within their grasp.

From the deployment of NATO forces in 2001 until their withdrawal in 2014, the Taliban came to learn the hard way they could never defeat the West in open combat. This is why they fell back on guerrilla tactics, planting roadside improvised explosive devices (IEDs) and resorting to suicide bombings.

But the Afghan government must not let its guard down, for it will only achieve its objectives by negotiating with boldness and conviction. Failing this, the Taliban are more than likely to give serious consideration to continuing the fight. If the Afghan forces and the country's Western allies maintain their resolve, the conflict is likely to trail off. Increasingly the Afghan Army and police, as they continue to grow in stature and professionalism, will become a more potent and effective fighting force. Western troops have remained to help the army and police in training and support functions well beyond the declared end to their role in combat operations in December 2014. We have every reason to be confident, but that confidence should translate into a preparedness concurrently to pursue a political solution.

It is going to require some bold initiatives to create a forum for reintegrating Taliban insurgents, meaning those who are not irreconcilable. We must think in terms of power-sharing deals with Taliban leaders and economic incentives for the rank-and-file militants. Afghanistan has been embroiled in conflict for more than three decades, everyone is weary of war and the Taliban know they are not supported by the vast majority of Afghans.

Concomitant with the ongoing military effort, it is crucial to convince the ordinary Afghans that we are not their enemy and that we can help them achieve their priorities. These people are not so different from us: their basic aspirations are a brighter future for their children, justice, decent infrastructure with clean drinking water and affordable electricity, for instance, and above all jobs and security. At the same time, the counter-insurgency strategy put forward by General David Petraeus, the former ISAF commander, must remain an integral piece of strategy. This is a political as well as a battlefield war and this embraces many aspects of nation-building, from education and reconstruction to tackling corruption and whatever else involves the local population.

I believe we have made many mistakes in the past and we've had to learn some harsh lessons. One of these is the fundamental need to invest in local solutions based on a much better understanding of what is tribally and culturally appropriate. The Taliban are virtually all from the Pashtun tribe, which comprises Afghanistan's dominant ethnic group. We must tread cautiously and not try to impose Western values on a highly traditional and largely rural society, even on matters we may find offensive, such as the status of women. Change will come in the longer term, but it will have to come from within, as we are already witnessing in the more cosmopolitan urban centres.

Jules Stewart's book is a graphic reminder of the errors that cost us so dearly in previous conflicts in Afghanistan. In the First Afghan War of 1838–42, we made the mistake of withdrawing the bulk of our troops to India once we took control of Kabul. We behaved like an army of occupation and made no effort to understand the Afghans' tribal sensitivities, much less their grievances. We reneged on our word by cutting the subsidies to the tribal chiefs to keep the roads safe. This was an affront they could not forgive and for which we paid a terrible price: the slaughter of 16,000 soldiers and civilians on the retreat to Jalalabad. Many of these blunders were repeated in the Second Afghan War of 1878–80. Believing the Afghans had been beaten into submission, we withdrew and forced the Afghans to accept a British Resident in Kabul to counter a perceived threat of Russian expansionism. Within three weeks the poorly-protected Resident and his entire staff had been murdered and the army was obliged to march back in to wage yet another bloody campaign. Britain walked away from the negotiating table in control of Afghanistan's foreign policy, a trophy we were persuaded to hand back after the Third Afghan War of 1919. Small wonder the Afghans, whose troops had been soundly trounced by the British Army and fledgling RAF, went home and declared victory. Stewart makes the point that Britain's three Afghan wars, above all the first one, demonstrate the need to act and negotiate from a position of strength, understand the Afghans' tribal customs and scrupulously abide by our word. The book is a vivid reminder of the follies committed in the past and which must not be repeated today.

The Afghan government must not fail in its effort to achieve national reconciliation. Its mission cannot fail because of the intoxicating effect this would have on those who oppose democracy. Nor should we abandon Afghanistan to its fate. We must not turn our backs on the country, for this would constitute a betrayal to the Afghan people, who were promised freedom, security and above all, peace. Worst of all, an abandonment of our Afghan allies at this juncture would risk weakening the government that is striving to rebuild its nation. Afghanistan's Thirty Years War must not be allowed to turn into a Hundred Years' War.

Acknowledgements

My thanks, first and foremost, go to General Sir David Richards, Chief of the Defence Staff, for taking time out from a devilishly busy life to provide a foreword to this book. I hope my efforts as a military historian prove worthy of this endorsement. I am also grateful to my agent, Duncan McAra, who with perseverance and encouragement opened the leaden doors of publishing to me when I first took the decision to turn from journalism to authorship. Helen Crisp, a practitioner in howler-spotting, has taken the trouble to scrutinise every word of my manuscripts, past and present, and for this I cannot thank her enough. I am indebted to my editor Joanna Godfrey for contributing her skills, and of course to I.B. Tauris for agreeing to publish the book – no trifling matter.

Introduction

In the 80-year period between 1839 and 1919, Great Britain fought three wars with Afghanistan. Britain's supreme interest in this country was to create a buffer against Russian expansionism. In the nineteenth century, the days of Great Game rivalry in Central Asia, the prevailing fear was that Russia would gain a foothold in Afghanistan, overwhelm her disorganized and poorly equipped army, or indeed strike a deal with whoever happened to occupy the throne of Kabul, to march Tsarist troops across the country's barren wastelands to the gates of India. From there, it was no great feat to break through the passes, notably the Khyber and Bolan, and send a Russian force pouring over the frontier to strike at the very heart of Britain's Indian Empire.

These fears were not only far-fetched, they bordered on the realm of fantasy. For the Russians, the Great Game meant harassment, not confrontation. Having large numbers of British troops occupied in defending India from an imaginary peril gave Russia a freer hand to operate in Europe and the Far East. This was particularly the case in the First Afghan War, when Russia managed to frighten Britain into dispatching a huge army into Afghanistan to depose a ruler who was falsely believed to be conniving with St Petersburg. The Army of the Indus launched its invasion after the supposed threat, Persia's Russian-backed siege of Herat, had evaporated. The tsar and his ministers had no appetite for a military clash with Britain in Asia. In the Second Afghan War, the Russians again taunted Britain by sending a military-diplomatic mission to Kabul, in spite of protests by the emir, who knew the British Government would take this as a provocation. As indeed it did. The Government rose to the bait by launching yet another invasion, a conflict that appeared to come to an end with an agreement to accept a British Resident in Kabul. Within three weeks the Resident and his entire staff had been murdered by a rampaging mob and the army once more found itself on the march into Afghanistan. The second stage of this war ended with a treaty that gave Britain control over Afghanistan's foreign relations. It was believed this step had finally eliminated the danger of Kabul forging unwelcome ties with foreign powers, namely Russia. Then in 1919, it was the Afghans who went on the offensive with an incursion into the Khyber Pass, touching off the Third Afghan War. This was the initiative of a weak emir who sought to exploit widespread civil unrest in India and appease the anti-British, fundamentalist elements of the Afghan military and clergy. Throughout this short-lived conflict, which in many respects could be considered a month-long skirmish, the hand of Russia was nowhere to be seen. The British scored a crushing defeat of the Afghan forces, thanks largely to the use of air power in support of superior ground troops. The emir, ignoring this development, sued for peace and proclaimed victory. He did so with some justification, for the Treaty of Rawalpindi was a tactical triumph for Kabul, inasmuch as the Afghans achieved their cherished objective, the restoration of full sovereignty.

It was now that the Russians made their appearance, and legitimately so, by being the first country to recognize the newly independent Afghan state. The Soviets, as they had now become, sent economic and military assistance to Afghanistan, thus stealing a march on their rival, the United States, which withheld diplomatic recognition for another 15 years. In 1979, Britain's 140-year-old nightmare came true when Soviet tanks rolled across the Oxus into Afghanistan, taking the Russians to those gates of India (now on the border of Pakistan), yet never showing any inclination to force them open. This was now of no great concern to England, for the British Empire had long since ceased to exist.

The Soviets failed to defeat the Afghan resistance, comprised overwhelmingly of the dominant Pashtun tribes of southern Afghanistan. But the British fared no better in a century of warfare with these fiercely independent people who inhabit both sides of the Durand Line, the boundary that was demarcated in 1893 between Afghanistan and British India. Throughout history every Central Asian conqueror had met with the same fate. For 2,000 years, a regular flow of invaders swept across Afghanistan. From Alexander the Great and Genghis Khan to Tamerlane and Babur, all the great conquerors of the ancient world were to discover, usually with a great deal of pain, that occupation of Afghanistan is not synonymous with subjugation.

The British found a worthy battlefield adversary in the Pashtun tribesmen of Afghanistan. The fighting was of a more savage nature than anything the troops had encountered in the years of Indian conquest. Here is an excerpt from the one of the letters home of Lance Corporal William Eaton of the 44th of Foot, who fought at the Battle of Mazina, a short, sharp engagement in 1880 during the Second Afghan War. This *Boy's Own Paper* five-hour battle took place after an eight-hour march in the blazing Afghan sun.

> We were ready for action directly we crossed a hill and we could see them about a mile off. We kept advancing and when we got within range we kept giving a volley into their midst. We had to force them at the point of the bayonet. I noticed one of the Afghans dressed in white and wielding a great long sword. Our captain saw him and he said to us, "Leave him to me, men, I'll settle him." A few minutes later I saw him engage in a desperate sword combat with the Afghan chief. Another Afghan was running with his sword uplifted to strike our captain in the back. There was not a moment to lose. I presented my rifle. I knew if I missed the Afghan would kill the captain. I took aim and the Afghan dropped dead, almost at the same time the captain dropped the Afghan chief. I now saw two more rush at the captain. He fought desperately with them. In a while one

of them cut the captain badly in the hand and he dropped his sword. In an instant he had his revolver out but it misfired. It would have gone badly with the captain, but by this time I had got up to him and I shot the one and ran my bayonet through the other.[1]

What makes the Afghans such an apparently truculent people, so vehemently hostile to outsiders, one might ask? The answer, of course, is that it all depends on the outsiders' intentions. If they come to conquer and plunder, it would be naive to expect them to behave otherwise. No one likes to have their country taken and one hopes it would not have been any different here if in 1940 the Germans had succeeded in invading England. Britain's great mistake was to treat Afghanistan exclusively as a military problem, ignoring from the earliest days that country's economic and social needs. The British went in wielding the sword and they met with the predictable response. The argument for a more enlightened approach would have involved a deeper, longer-term commitment, with the usual host of civil servants, educators, law makers, engineers and the like following in the footsteps of the military. In a word, the type of benevolent colonialism that developed over the many years of British rule in India. It is unhelpful when a British Cabinet minister dismisses Afghanistan as 'a broken thirteenth-century country', implying that the obligation of advanced societies is merely to keep the lid on a rogue state. It is also a rather ironic statement, for in the thirteenth century Islamic scholars, scientists and artists represented almost the only light in a world of darkness.

Colonialism has its evils, to be sure, but it is worth noting that every year the Government of India submitted an account to parliament, quite unreadable and usually running to more than a thousand pages, entitled *Report on the Moral and Material Progress of India*. The Government was obliged to state what it had achieved in areas like combating disease, alleviating poverty and hunger and generally improving the lives of its subjects. The creation of a vast, unwieldy and corrupt bureaucracy was nevertheless, along with the English language, one of Britain's most precious gifts to India. This writer was in Pakistan, also an inheritor of a British-imposed infrastructure, in 1988 when the country's military dictator General Mohammed Zia-ul-Haq was killed in an air crash in suspicious circumstances. In spite of the country's notorious reputation for volatility, there was no implosion, no breakdown of law and order, no chaos in the streets. The bureaucracy continued to function, as did the airports, the railways and all social services. It is arguable, albeit unfashionably so, that if India had not been colonized by a European power, the subcontinent might today resemble a vast, ungovernable Afghanistan.

This raises the issue of the Fourth Afghan War and whether the West, by yet again committing immense military resources to Afghanistan, is ignoring the lessons of history. The answer is that the result of the current war in Afghanistan will depend on our ability to implement an effective strategy to provide security, good governance, economic opportunity and decent infrastructure for the country's long-suffering people. This is an approach which, until it was too late, Britain never took into consideration, not when its army was in occupation of the country nor during the century of British dominion of the North-West Frontier, the tribal belt that for all intents and purposes is a linguistic and ethnic extension of Afghanistan. If the Western armies now deployed in Afghanistan, and realistically this means the USA and Britain, fail to make this effort to win the proverbial hearts and minds, the invasion launched in 2001 will inevitably end in disaster and humiliation. This is the lesson of history.

CHAPTER 1

Cradle of Political Insanity

The Honourable Mountstuart Elphinstone rode through the gates of Peshawar on a dazzling spring morning in 1809, at the head of a column of 450 infantry and cavalry with drums and trumpets playing, behind which trailed a mile-long multitude of Indian servants, 600 roaring baggage camels and a dozen elephants.

The city's Pashtun inhabitants poured out by the thousands from their market stalls and mud-walled houses, and even the fabled Storytellers' Bazaar was left deserted, as the tribesmen fought to catch a glimpse, many for the first time, of the fair-skinned sahibs in their fine blue cavalry uniforms glittering with gold buttons and braid. The banks on each side of the road were covered with people and many climbed trees to watch the procession. The mass of people swelled as Elphinstone and his entourage reached the city centre but, as he noted, this caused no inconvenience: 'The King's Horse that had come out to meet us charged the crowd vigorously, and used their whips without the least compunction.'[1] One of the Afghan troopers in particular attracted Elphinstone's attention. He was a ferocious-looking soldier known as Rusool the Mad, an officer in the King's Guard who had acquired, and perhaps it is best not to inquire by what means, an English helmet and cavalry uniform. He carried a long spear, bellowing with a loud and deep voice as he charged the tightly-packed throng at speed. 'He not only dispersed the mob', Elphinstone recalls, 'but rode at people sitting on terraces with the greatest fury.'[2]

Such was the tumultuous finale to a four-month trek that had begun in Delhi, where Elphinstone was handed his marching orders by Lord Minto, the Francophobe governor general of British India, to proceed westward to explore the uncharted land of Afghanistan. Elphinstone, who had just turned 30, was a colonial administrator of superb intellect in the tradition of the soldier-scholars who engendered so much interest at home in India's cultural as well as its pecuniary treasures. He was a fluent Persian speaker and became a noted authority on Indian literature and philosophy. Elphinstone was posted to Benares, a key centre of British intelligence gathering, where he was to put to good use his talents for picking up useful gossip, from the bustling bazaars to princely gatherings. His achievements were not confined to scholarly pursuits. After his arrival in Calcutta, he was appointed to the Duke of Wellington's staff and distinguished himself at the Battle of Assaye, which definitively crushed the Maratha threat to British dominion in India. He was certainly a more colourful character than his fellow Scotsman Gilbert Elliot, the future Lord Minto. Despite remaining a lifelong bachelor, Elphinstone declared himself fond of 'philandering' with Calcutta society ladies and Indian dancing girls.

Minto, on the other hand, was an able administrator and a well-intentioned, though less engaging personality who laboured under an exaggerated fear of Napoleon's hostile designs on India. He had been educated at the Pension Militaire in Fontainebleau, an experience that did little to endear him to his host country. Minto was a vigorous opponent of the French Revolution who spent much of his early diplomatic career in Italy and later in Austria, galvanizing local forces to oppose Napoleon's armies. At the time of Elphinstone's mission, Britain had been at war with France for five years. The imperial struggle for mastery in Europe had another six years to run before Napoleon's final defeat at Waterloo, the battle in which another Elphinstone, Mountstuart's first cousin William, was to see active service. He was later to go on to become one of the chief culprits behind Britain's worst military catastrophe in Afghanistan.

In 1808, the Government of India had received word via

agents in Tehran that the French were intending to carry their war with Britain into Asia. There were reports of Napoleon conniving with Russia's Tsar Paul I to launch a joint invasion of India, though it was never clear how the spoils were to be divided up, namely if the invading armies were to defeat Britain, whether India was to be a French or a Russian Raj. These fears were founded on reasonably accurate intelligence, for in 1801 the two imperial rulers agreed an alliance to bring their scheme to fruition. The plan was for the Russians to sweep down across the Hindu Kush and Afghanistan, while the French would march their army through Persia to link up with the tsar's forces on the Indus. The plan collapsed quite suddenly with the tsar's assassination later that year, but Napoleon continued to dream of Eastern conquest. The signing of the Treaty of Tilsit in 1807, which ended hostilities between the two imperial powers, brought France and Russia together in an alliance that rendered the rest of Continental Europe almost powerless. When mapping out the strategy for his Egyptian expedition, Napoleon had in fact planned to subsequently join forces with Indian princes to attack the British. But when Bonaparte was forced to beat a retreat from the Middle East, putting a French army across the Indus had ceased to be a viable undertaking, even before the Elphinstone mission set off for Afghanistan.

Elphinstone's task was to negotiate a treaty with the Emir Shah Shuja ul Mulk and secure undertakings to resist an advance across his territory by any foreign power. The Afghan monarch was deemed to be an arrogant ruler who was strongly opposed to receiving a British mission. For this reason the Government thought it appropriate to send a deputation in a style of great magnificence, in order to duly impress the king with British might. Shuja's Sikh friends, never ones to shirk from an opportunity to stir up trouble for the British, had warned the emir he would be coerced into giving up territory to British India. But being a vain and covetous man, Shuja eventually relented and sent word that Elphinstone would be permitted to proceed to Peshawar, though keeping him at arm's length from the Afghan heartland. Shuja had heard tales of Britain's fabulous wealth and he anticipated receiving handsome gifts from his European visitors.

His European visitors, meanwhile, were toiling with great hardship across the Thar Desert. The column often marched by night to avoid the consuming heat of day, drums beating to alert stragglers who got separated from the main body. Men deserted in droves, such was their terror of the waterless expanses and the ever-present danger of cut-throat bandits who swooped down on the defenceless camp followers. As they neared the Indus, Afghan horsemen from the wild border hills beyond began to make their appearance. These warriors were equally mystified by the approaching horde, taking them for Hindus or Mughals. 'They believed we carried great guns ... and that we had certain small boxes, so contrived as to explode and kill half a dozen men each, without hurting us', wrote Elphinstone. 'Some thought we could raise the dead.'[3] Once ferried across the Indus, Elphinstone found himself in tribal territory proper, amongst the Pashtuns of today's North-West Frontier Province of Pakistan. The mission had been forced to trace a southerly route instead of following a direct line westward. The shorter journey would have taken the mission through Sikh territory, which was under the iron-fisted rule of the diminutive one-eyed Maharajah Ranjit Singh. Though an uneasy truce prevailed between the British Government and the King of the Punjab, Minto was eager to avoid provoking the Sikhs, who had a two-to-one superiority in artillery.

Shuja prepared a royal residence to receive his European guests. He sent provisions for 2,000 men, most of which was discarded and left to rot, and only with great difficulty was Elphinstone able to persuade the emir to restrain his cordiality. In his first audience with Shuja in Peshawar's Bala Hissar fortress, Elphinstone lavished presents of great splendour on the king. Shuja was duly impressed, even for one who wore on his arm the fabled Koh-i-Noor diamond, one of history's most romantic gems, which eventually became part of the British Crown Jewels. Back in Calcutta, Minto was less impressed by the mission's lavish expenditure, which to his mind far exceeded the necessity of the occasion. The emir was especially delighted with a pair of magnificent custom-made pistols and he was also pleased to receive an organ, though we have no record of how it was put to use. On returning to his residence after the first meeting with Shuja, Elphinstone was delivered a royal request for yet another gift, a pair of silk stockings like the ones worn by the British visitors.

Minto's misgivings were well founded when measured against the results achieved by the mission. Elphinstone brought back a wealth of scholarly information on Afghan geography, tribal customs and culture. But on the diplomatic front, it took three months for him to negotiate 'a somewhat useless treaty, by the terms of which no Frenchman or other European was to be allowed to enter Afghanistan'.[4] Before the year was out, Shuja had been ousted from the throne of Kabul by his elder brother Mahmud. The former emir wandered alone as a fugitive across Afghanistan. He was seized by Ata Mohammed Khan, the son of his former vizier, or chief minister, who subjected the fallen monarch to humiliating indignities. Shuja was confined in a fortress in Attock, a lancet was threateningly held over his eyes, he was nearly drowned in the Indus with his arms bound, and when released from his tortures, he travelled to the Punjab to take refuge with Ranjit Singh in Lahore. The wily Sikh turned out to be less of a loyal ally than Shuja had envisaged. The emir was made a virtual prisoner. He was dispossessed of the Koh-i-Noor diamond and only succeeded in making an ignominious escape by creeping through the city sewer in disguise. Shuja's final mortification was to seek asylum in Ludhiana from the same British Government that he had entertained in such splendour in Peshawar.

Thirty years were to pass before Britain dispatched another expedition to Afghanistan. On this occasion, the British embassy was sent not to negotiate but to impose its will by force of arms. The intervening three decades of expansion were characterized by 'a sequence of tactical decisions made in response to local and sometimes unexpected crises'.[5] It was George Douglas

Campbell, the Duke of Argyll, who recognized in the heyday of Britain's imperial expansion that 'The pure Instinct of Dominion, unadulterated by another other feeling more rational than itself, is one of the very strongest of human passions.'[6] For Argyll, that instinct has always been strongest with those who fielded the strongest armies, and through them it has been the most powerful of impulses in the history of human progress, never more manifest than in the British conquest of India.

The years 1810 to 1818 saw bits of territory in India and the Far East systematically gobbled up by a colonial power determined to seize the initiative from its French, Portuguese and Dutch rivals. The British captured Goa, taking advantage of the French occupation of Portugal. Nor did France escape the British Empire's voracious clutches. Réunion and Mauritius were taken, and then Minto turned his attention to the Dutch possessions in the East Indies. Amboyna and the Moluccas fell, followed by Java, in an expedition in which Minto himself took part. In this period, the East India Company's army concluded the last of three wars with the Maratha Confederacy.[7] In 1818, Minto's successor, the Marquess of Hastings (not to be confused with Warren Hastings, who served as governor general from 1774 to 1785), scored a victory over the Marathas at the decisive Battle of Konegaon, which left the British with dominion over most of what was to become the Indian Empire south of the Sutlej River. There remained the question of the Sikh kingdom of the Punjab and the tribal lands to the north-west, to be dealt with at a later date.

With the final pacification of the Maratha Confederacy, the Government's attention was drawn to events taking place beyond India's frontiers. If France was no longer seen as a threat to British hegemony in India, the shadow of Russian expansionism loomed large in the strategic thinking of Calcutta and Whitehall.

The fear of a Russian attack against India created a political fault line that threw the two great imperial powers into conflict for supremacy in Central Asia. The 'Great Game' was a phrase coined by Captain Arthur Conolly, a British intelligence officer who in 1842 fell foul of the deranged Emir of Bokhara and, along with his fellow officer Lieutenant Colonel Charles Stoddart, was publicly beheaded in that city on charges of spying.

The Great Game was a nineteenth-century forerunner of the Cold War, a battle involving secret agents and diplomats, in which not a shot was fired between British and Russian forces. This confrontation was played out roughly between the settlement of the Russo-Persian War of 1813 and the Anglo-Russian Convention of 1907, which left Persia carved up into spheres of Russian and British influence. This treaty officially marked the end of the Great Game, though Russia kept the diplomatic heat turned up until Britain's departure from India in 1947. Russia always coveted India and its warm-water outlet to the Arabian Sea. Around the time the 1907 accord was being negotiated, by which both powers were expected to retreat from their century-long war footing, the influential Russian Army general and Orientalist Andrei Evgenevich Snesarev wrote:

> England must not dictate our path to us. Rather, we should direct her fate. Above all, remember that . . . beyond the snow-capped mountain range of the eastern Hindu Kush lies India, the foundation of British power, and perhaps the political key to the whole world.[8]

It is no coincidence that Persia takes centre stage from start to finish in the Great Game drama. Napoleon believed he could conquer India with an army of 50,000 men marching through Persia. This fantasy aside, the spectre of Persia serving as an invasion route into India was taken seriously enough for the government to negotiate a treaty with the shah, bringing Persia into a defensive alliance by which it would oppose any European force crossing its territory towards India. The treaty was given definitive form five years later. The final agreement provided for a frontier settlement to be negotiated with British assistance. Likewise, the shah agreed to do his best to prevent Central Asian (a euphemism for 'Russian') neighbours from attacking India from Persia.

Persia and India of course do not share a border. The real cause for alarm was of Persia as a gateway into British India's buffer state: Afghanistan. The Royal Navy was in undisputed control of the sea lanes, and the great Himalayan and Hindu Kush ranges rose as a natural barrier to any army foolish enough to attempt a crossing. Therefore the only viable attack could come from the west, across the unprotected passes that straddle the present Pakistan-Afghan frontier.

Despite his misplaced anxiety over the French peril, Minto, the first of six Scots who were to rule India, was 'a discerning politician and diplomat and an energetic administrator'[9] who set a sound and useful example to his successors in office. Unfortunately, Minto's nephew Lord Auckland landed at Bombay in 1835, destined as governor general to chart a downward course from his prior successes as President of the East India Company's Board of Trade and First Lord of the Admiralty. Another candidate, Lord Heytesbury, a nobleman of proven diplomatic skill, had been chosen before by Robert Peel's Conservative Government and had actually taken the oath of office. 'Lord Palmerston's opposition to Heytesbury seems to have grown out of the fact that he had been Ambassador to Russia, and was known to be an admirer of Tsar Nicholas.'[10]

Auckland arrived at a time of prevailing tranquillity in India. He was a confirmed bachelor who left his social engagements in the hands of his two adoring sisters. The letters of one of them, Emily Eden, offer a unique insight into the lively frivolity of British India's courtly society. Here's a charming sample:

> There is a Colonel E. Come into camp today . . . He is about the largest man I ever saw, and always brings his own chair with him, because he cannot fit into any other. [. . .] To finish off Colonel E., I must mention that the officer who commands his escort is called Snook, and that his godfathers, to make it worse, called him Violet. He is a little man, about five feet high, and is supposed to have called out to three people for calling him Snooks instead of *Snook*.[11]

Auckland was too eager to emulate his uncle's hard-line policies on the foreign threat to India. His bogeyman was Russia and in this endeavour there was no shortage of supporters in India as well as at home. Auckland fell victim to an acute case of Russophobia and this brought on 'a crisis which probably need not have occurred, and which he did not show any great talent in dealing with. He was one of the least distinguished of those who have held his place.'[12] Variously described as 'a notorious and useless jobber' and 'a man without shining qualities', Auckland imagined invasion by an enemy in retreat and betrayal by a monarch whose hand was extended in friendship.

Almost inevitably, behind such a mediocre administrator of empire there stood a figure of Machiavellian political astuteness, combined with the ambition of a Wall Street banker. This was the Indian-born William Hay Macnaghten, political secretary to the governor general. Of Auckland's three closest advisers – Henry Torrens, John Colvin and Macnaghten – the latter was the most fanatical Russophobe and he could also be assured of the governor general's ear on matters of imperial security.

Macnaghten was unquestionably the most refined specimen that his clan had produced for decades. One is hard-pressed to envisage a blood relationship between the Sanskrit- and Arabic-speaking bespectacled scholar, and his forbear John Macnaghten, who had survived a botched execution for murder, only to clamber back up the gibbet demanding the job be done 'properly'. Macnaghten's linguistic ability was such that he was reputed to speak Persian with greater fluency than his native English. Indeed, his last mortal words, a plea for mercy from his assassin, were spoken in that language.

Auckland and his sisters luxuriated in the social gaieties and bracing mountain air of Simla, the summer hill station of the Raj which one contemporary historian described as 'That pleasant hill sanatorium ... where our Governors General, surrounded by irresponsible advisers, settle the destinies of empires, and which has been the cradle of more political insanity than any place within the limits of Hindustan.'[13] Auckland was shortly to validate the accuracy of this statement. While the governor general and his entourage delighted in tea parties, balls and lawn tennis, some 500 miles to the west events were unfolding that were to eventually bring about Auckland's downfall in disgrace.

Lord Palmerston was one of the first in Government to react to the gathering war clouds over Herat, Afghanistan's most westerly city, which in 1835 was coming under the threat of Persian attack. In a dispatch to Henry Ellis, Britain's Minister to Persia, the hawkish foreign secretary stated with characteristic bluntness: 'You will especially warn the Persian Government against allowing themselves to be pushed on to make war against the Afghans.'[14] The key phrase here is 'to be pushed', for Palmerston clearly spotted the trouble-making hand of Russia behind the shah's war preparations.

The Persian ruler Shah Mohammed was more than justified in rattling his sabre at the Afghans. Shah Kamran, the Governor of Herat, routinely allowed Turkmen raiders to round up Persian subjects, who were herded to the slave markets of Central Asia. To be sure, Kamran was not the most engaging of characters. 'After a notable career of criminal self-indulgence', writes one historian, '[he] was enfeebled by age and debauchery. Pock-marked, no longer able to satisfy his sexual appetites, he was now addicted to the bottle.'[15] Kamran was, however, able to indulge his penchant for cruelty and he regularly had his Shia Muslim citizens,[16] who were considered untrustworthy, tortured, butchered and sold into slavery. The centre of the slave trade was Khiva, whose bazaar held tens of thousands of these hapless souls, with nothing to look forward to but a life of forced hard labour, or having their ears nailed to the city gates if they attempted to escape. In one of his own periodic slave raids, Kamran had encroached on Persia's eastern border region of Seistan, in flagrant violation of an 1834 mutual non-intervention treaty made between Herat and Persia. Two years later he swooped on Persian Khoristan and carried off 12,000 persons to be sold as slaves.

When the shah acceded to the throne in that year, his country was totally under the influence of Russia, to whom Persia had lost two wars and a considerable amount of territory. After being dealt a second crushing defeat in 1828, Persia was obliged to pay a large sum of money to the tsar for having inconvenienced his army.

In order to free their hand in Persia's coming conflict with Herat, the British Government agreed to extend the money the shah desperately needed to meet the first instalment of his Russian obligations. With good reason Britain had earned its reputation as Europe's paramount wheeler-dealer: the arrangement was that in exchange for making good on Persia's debts, Britain would be released from Clause VIII of the 1814 treaty that bound Calcutta to refrain from interfering in any war between Persia and Afghanistan. In other words, should Herat fall, Britain was free to step in and oppose a Persian advance through Afghanistan, which would be interpreted as a threat to India. Not that Palmerston would consider himself restrained by such diplomatic minutiae if it came to the crunch, but the partial abrogation of the treaty sent out a clear message to the Russians that Britain meant business.

In December 1835 the shah formally announced his intention of launching a military campaign against Herat the following spring. Shah Mohammed proclaimed, with reason, that there was no such thing as an Afghan government or consolidated state, and with somewhat less justification that a large portion of Afghanistan belonged to Persia. A fortnight later, Ellis in Tehran dropped the bombshell on an already extremely irritated Palmerston, reporting that 'the Russian Minister at this Court [Count Ivan Simonich] has expressed himself in very strong terms, respecting the expediency of the Shah losing no time in undertaking the expedition against Herat'.[17]

The shah set out for Herat with a small detachment of cavalry in June 1837, gathering as he rode across the dusty plains units of the Persian Army, which numbered more than 30,000 men by the time they reached the border at Torbat-e-Jam. The army resembled more a noisy rabble than a disciplined fighting force. The 650-mile march had taken twice the normal journey time, even as a heavily laden column, and it was late November before the Persians stood before the city ramparts. The Afghan defenders were distinctly unruffled by the ragtag horde encamped below the walls, to the extent that most of the city's five gates remained open and caravans from Bokhara continued to file in with normality. Shah Mohammed's strategic planning was not aided by the presence of Russian and British agents who rode with the army. Each kept an anxious eye on the other while secretly competing for the shah's ear, the Russians pushing to get on with the attack, the British seeking to mediate a peace settlement.

The British mission was headed by a well-known Russophobe, John McNeill, a Scottish diplomatist and surgeon whom Palmerston had appointed to replace Ellis in May 1836. McNeill relied on an extensive network of native spies, euphemistically known as 'newswriters', to provide intelligence on movements within the Persian camp, and even to pass on the contents of official correspondence between the shah and his Russian allies. 'Native channels of information not unworthy of credit' was a phrase typically inserted in a dispatch to signal it contained reliable intelligence. Early in the siege, McNeill's reports spoke of a half-starved Persian force lacking in discipline, suffering from bitterly cold weather, short of supplies and without any preparations in place for a counterattack by the Herat defenders. The Persians far outgunned the Heratis and had there been any skilled artillerymen in their ranks, there is little doubt the city would have fallen within weeks. Their shots, however, seemed to land everywhere but on the city's embattlements, thus emboldening the garrison to mount cavalry forays against the Persian trenches. Not that the Herati soldiery distinguished themselves as battlefield tacticians: in one sortie the rear attacked the head of its own column, and on the subject of heads, the Heratis would consistently break off an attack to collect the heads of fallen Persians, which were paraded into the city on lances and offered up for a bounty. It was not unknown for the troops to bring in the severed heads of their own comrades, disguised as those of Persian soldiers.

At this juncture there appeared in Herat one of those eccentric Victorian adventurers whose exploits made the Great Game a household name for high romance. So masterful was Eldred Pottinger's covert career that Sir Henry Lawrence, one of the paladins of the North-West Frontier, would later write of him that 'India, fertile in heroes, has shown since the days of Clive no man of greater and earlier promise than Eldred Pottinger. Yet, hero as he was, you might have sat for weeks beside him at table and not have discovered that he had seen a shot fired.'[18] In a word, an undercover agent of the highest order.

It could arguably have been Pottinger's Irish pluck that took him, alone, disguised as an Indian horse dealer and unmolested, across the bandit-ridden Khyber Pass to Kabul. Born in Ireland in 1811, Pottinger's brief life covered a military career that began with enlistment in the Bombay Artillery at the age of 16, rising to the rank of major when he was appointed Envoy in Kabul during the Afghan uprising against the British garrison. Pottinger safely reached the court of the Emir Dost Mohammed[19] in August 1837, and then decided to carry on to Herat. After a brief sojourn at the emir's court, he sensed that the Afghans were on the verge of seeing through his ruse, so he slipped out of Kabul by night, thereafter adopting the identity of a *syed* pilgrim, or holy man directly descended from the Prophet Mohammed. On his way across Hazara territory, Pottinger was taken prisoner at the fort of Yakub Beg, a notorious slave trader. He was subjected to a lengthy interrogation, and Pottinger was convinced the game was up the moment his captors went through his baggage and discovered a copy of Elphinstone's account of his journey to Afghanistan, along with some pencils, a compass and a Persian-Pashtu dictionary – an odd collection of accoutrements for a holy man on pilgrimage. Pottinger had to quickly summon up all his Irish wit to talk his way out of a very tight spot. He recorded the following in his diary:

> Now that I looked back, well knowing the imposition I had been practising, I could not conceal from myself the extent of the case, and that a discovery had really been made, but that hitherto good fortune had saved us. For the barbarians were not

certain in their own minds, though a grain more evidence or the speech of a bold man would probably have decided the affair.[20]

After several days of nerve-racking grilling, encircled by Yakub Beg and his mullahs, the Hazaras lost interest in this bizarre traveller and Pottinger was sent on his way. He reached the gates of Herat with no further mishaps nearly a month after departing Kabul.

In Herat, Pottinger soon managed to arrange a meeting with the real power behind the throne, the vizier Yar Mohammed, by all accounts a thoroughly objectionable character. 'Of all the unscrupulous miscreants in Central Asia', Kaye tells us, 'Yar Mohammed was the most unscrupulous. His avarice and his ambition knew no bounds, and nothing was suffered to stand in the way of their gratification. Utterly without tenderness or compassion, he had no regard for the sufferings of others.'[21] Yar Mohammed had a face to match his loathsome personality, with a stern countenance, an overhanging brow and an abruptly receding forehead. Confronted with such an unsavoury specimen, Pottinger's first thought was to make a discreet exit from Herat. But this would mean leaving the city empty-handed, with no useful intelligence on the Heratis' preparations to meet the approaching Persian force. The jury is out on whether Pottinger was acting as a secret agent, or had simply travelled to Herat to fulfil his lust for adventure. The fact that he chose to remain in the city, and moreover to risk his neck by revealing his true identity to Yar Mohammed, suggests that he was indeed on a mission of sorts. The British authorities were certainly aware of Pottinger's presence in Herat and the accolades, as well as his promotion to the rank of major, that followed once the siege was eventually lifted, tend to substantiate the view that this journey was not undertaken solely to satisfy Pottinger's intellectual curiosity. McNeill and Auckland lavished flowery praise on the young Irish officer, who was hailed as the mastermind of Herat's salvation.

On 17 September, Shah Kamran with his army returned to Herat from campaigning abroad. That morning, Pottinger mingled with the thousands who lined the streets to watch the entry of the royal procession of several hundred baggage animals, executioners holding aloft their bloodied swords, a throng of eunuchs and a motley crowd of foot soldiers tramping behind the cavalry. Many of these onlookers, the commoners of the city's Shia sect, secretly longed for the Persians to deliver them from the tyrannical clutches of the Sunni ruling class.

It was at this point that Pottinger took the plunge by making his identity known to Yar Mohammed. He took this decision after having seen with his own eyes that Herat was the logical staging point for an invasion of British India. The surrounding countryside was rich in mineral resources and farmland for equipping and keeping supplied a large fighting force on the march, whose ranks could be augmented by an abundance of local manpower.

The crafty vizier at once saw that Pottinger's British Army experience and Government connections might be put to good use. Pottinger was invited to take on the role of strategic adviser in the defence of Herat. When hostilities began, one of his first efforts was to organize the cavalry sorties along professional lines, only to find himself frustrated by the persistent lack of discipline in the ranks. Pottinger could do little to organize an effective offensive strategy, therefore most of his efforts were concerned with keeping the Persians at bay outside the ramparts, as well as acting as mediator between the shah and Kamran. As the weeks dragged on and with food shortages beginning to take their toll in the city, Pottinger feared Herat's imminent collapse. He also began to suspect the duplicitous Yar Mohammed of plotting to defect to the enemy camp, a not uncommon practice amongst Afghans who will spare no effort to join up with whoever has the upper hand.

In early February, Pottinger left the city and slipped through the Persian lines to seek an audience with the shah. When he entered the Persian lines, Pottinger was first taken to the Russian general commanding the approaches to the shah's tents. This lifted any doubts as to who was literally calling the shots in the Persian camp. Pottinger learned that Russian officers and troops were in fact engaging in open combat with the Heratis. The mission ended in failure, for Pottinger knew the Herati chieftains would never accept the humiliating ceasefire terms set out by Shah Mohammed, which included a demand that Yar Mohammed and Shah Kamran tender their submission by kissing the feet of the 'King of Kings'.

A disappointed and angry Pottinger rode back through the city gates to take up his post on the ramparts. The campaign came to a sudden climax in July 1838, when the shah launched a surprise massive assault on the battlements. Pottinger was on hand to help repel the attack, which he declared was being pushed forward with an energy that had not until then marked the campaign. The situation took a desperate turn when the Persian advance guard, attacking with fury, forced a breach in the inner walls, which were held to be impregnable. To his horror, Pottinger found the garrison in full retreat before the advancing Persian soldiers. Worse still, he spotted Yar Mohammed crouched under the ramparts in a miserable state of despair. Pottinger struggled in vain for several minutes to rally the vizier to action. He finally had to drag him to the breach, where the Afghans could see the chief minister supposedly standing fast and urging them to follow his leadership. As if startled out of a daydream, Yar Mohammed suddenly rose to the occasion, raising his staff above his head and shouting to his men to drive the Persians from the walls. 'Had Yar Mohammed not been roused out of the paralysis that had descended upon him, Herat would have been carried by assault', writes Kaye. 'But the indomitable courage of Eldred Pottinger saved the beleaguered city. The Persians, seized with a panic, abandoned their position and fled. The crisis was over, and Herat was saved.'[22]

Why did the shah choose this particular moment, eight months into a stalemated siege, to embark on an all-or-nothing onslaught? The answer is to be found in the diplomatic war that was being waged behind the battlefield. The action shifts from Herat to London, where Palmerston was in a rage over reports of Russia's flagrant intervention on the Persian side, as well as the tsarist Government's machinations in Kabul at the court of Dost Mohammed. The political manoeuvrings could hardly have been more convoluted. The drama unfolded as follows.

Dost Mohammed proclaimed himself a steadfast friend of England, yet he was so obsessed with the loss of Peshawar, which had been conquered by the Sikhs in 1822, that he was prepared to offer himself and his kingdom to the Persians in exchange for military aid to recover the sacred city of the Afghans. The British needed to keep Dost Mohammed in their camp to take the brunt of the feared advance by a Persian-Russian force. Auckland, on the other hand, had no intention of alienating Ranjit Singh, whose mighty army, should the Afghans fail to hold off an invader, would form the first line of defence on the Indus. In any case, the Sikhs and their much coveted territory of the Punjab was a matter to be dealt with at a later date. Dost Mohammed's half-brothers, the powerful sirdars, or rulers, of the Pashtun heartland of Kandahar, were Sunni Muslims. As such they would ordinarily have regarded their sworn enemies, the Persians of the Shia sect, with utter contempt. Yet both parties had concluded a treaty by which the sirdars were to provide 12,000 cavalry and 12 guns, to march with the Persians and be given suzerainty of Herat when the city was taken. This unholy alliance was of grave concern for the Heratis, but more so for the British, for the treaty had been drafted by none other than the Russian ambassador Simonich. When news of this pact reached Dost Mohammed, it left the emir in a state of despondency. If his Kandahar brothers had signed a protocol with the Shah of Persia behind his back, his throne was now threatened from the west, thus compounding the ever-present menace of a hostile Sikh army on his eastern border. The East India Company's army was the logical ally to turn to for help, but Auckland had rejected Dost Mohammed's pleas for assistance. The only power that came forth to offer the emir the protection he desperately sought was Russia.

Palmerston was incensed not only by Russia's mischief at Herat; the foreign secretary was equally alarmed over a frustrating turn of events at Kabul, where a diplomatic tug-of-war between British and Russian agents had ended in defeat for the Government. In 1837, Kabul had become the epicentre of Great Game intrigue. Auckland had chosen a glittering society darling of the day to head a 'commercial' mission to the Afghan capital. Alexander (later 'Colonel Sir') Burnes had earned a place of honour in London literary salons. The young – and yet again – Scottish adventurer was known by the dashing sobriquet of 'Bokhara Burnes' and his travel narrative was a runaway best-seller that sold 900 copies on the first day. Burnes had managed to enter the closed city of Bokhara and on his return to England had even spent a few hours with William IV, briefing the king on his astonishing journey. Auckland had informed Dost Mohammed that he intended to send a British agent to Kabul as a commercial representative to discuss ways of opening Afghanistan to trade. This was at best a half-truth, one of the governor general's many false utterances that were in due course to lead Britain into war. Burnes's real brief, quite simply, was to secure Dost Mohammed's loyalty in the event of a Russian push through Afghan territory.

Burnes was received in Kabul in September 1837 in great pomp and splendour, seated on an elephant alongside Akbar Khan, Dost Mohammed's favourite son. This was Burnes's second visit to Kabul. The first had taken place in 1831, during his celebrated voyage across Central Asia, disguised variously as an Armenian, a pilgrim and a merchant. A fluent Persian speaker, a man of boundless self-confidence and diplomatic astuteness, Burnes was in every respect the ideal emissary to undertake the mission to Kabul. In every respect, that is, but one: he was an admirer of Dost Mohammed and firmly believed in the emir's desire to remain steadfast to the British cause. In this, Burnes was to find himself in a minority of one. Auckland had no personal knowledge of Afghanistan. The governor general's thinking on this subject was moulded by a coterie of influential advisers. Macnaghten and McNeill have already been cited as inveterate Russophobes. Another of Auckland's close confidants was Sir Claude Wade, the son of an Irish Bengal Army officer, who was a political agent at Ludhiana. He was a short, obese man with a fondness for eating and sleeping who, when not indulging his sybaritic pastimes, spent his time looking after British interests at the court of Ranjit Singh. He also had as his charge the exiled Shuja, on whose behalf Wade was lobbying hard to have reinstated on the throne of Kabul by any means possible.

Dost Mohammed had a great deal of personal sympathy for Burnes, though he was slightly put off by the paucity of gifts – a pistol and a telescope, along with an assortment of pins and needles for the ladies of the zenana – brought to him by the envoy of the all-powerful British Government. This was a far cry from the array of finery Elphinstone had bestowed on Shuja. However, Burnes had been sent an official dispatch stipulating the gifts 'ought not to be of a costly nature, but should be chosen with a view to exhibit the superiority of British manufacturers'.[23]

In one of Burnes's first letters to Macnaghten, he alerts the political secretary to an alliance being forged between the Kandahar sirdars and the Shah of Persia. Burnes then goes on to reassure Macnaghten that Dost Mohammed is equally disturbed by this turn of events, which according to the emir would result in the fall of Herat and probably Kandahar as well, 'if some arrangement was not speedily devised to check her (Persia's) approach, as it was very clear from every account that Persia could not of herself act in this manner, she must be assisted by Russia'.[24] Burnes's pleas for Government support for Dost Mohammed as a trustworthy bulwark against Russian aggression fell on deaf ears. A few days after his first audience with Dost Mohammed, Burnes was singing the emir's praises to Macnaghten. He assured him that 'Dost Mohammed will set forth no extravagant pretensions, and will act in such a manner as will enable the British Government to show its interest in his behalf and, at the same time, preserve for us the valued friendship of the Sikh chief.'[25] Here was a promise of loyalty from Dost Mohammed, who only asked the Government to meet him half-way, by acting as honest broker to settle the dispute with Ranjit Singh over Peshawar. But Auckland, or rather Macnaghten, was bent on a different course.

The governor general sent a letter to the emir bluntly telling him to give up any pretentions of recovering Peshawar. In this missive Dost Mohammed is cast as a misguided man, while Ranjit Singh shines as the Government's valued ally. Auckland wrote:

> In regard to Peshawar, truth compels me to urge strongly on you to relinquish the idea of obtaining the government of that territory. From the generosity of his nature, and his regard for his old alliance with the British Government, Maharajah Ranjit Singh has acceded to my wish for the cessation of strife and the promotion of tranquillity, if you should behave in a less mistaken manner towards him.[26]

(Less than a decade was to pass before Britain waged two wars against its former Sikh allies, the outcome of which was the annexation of the Punjab to the Indian Empire.) Auckland went on in a rather coarse style to caution that Burnes would be withdrawn from Kabul if Dost Mohammed entertained any thoughts of turning to the Russians for assistance. He also threatened to call off all attempts to intercede on the emir's behalf with the Sikhs, whatever that was intended to mean.

Dost Mohammed had in fact himself written to Ranjit Singh seeking a compromise arrangement to the Peshawar imbroglio. The Afghan king's proposal in essence called for a member of the Barakzai family to govern Peshawar, while leaving the city under Sikh sovereignty. Wade saw to it that this letter never reached Ranjit Singh's hands. He instead had it delivered to Auckland, whence the governor general's arrogant dismissal of Dost Mohammed's hopes of regaining Peshawar.

Alexander Burnes was not the only foreign agent lurking about the bazaars and listening posts of Kabul in late 1837. It was Burnes, in fact, who first alerted the Government to the appearance of another visitor from abroad, in a memo that set Macnaghten's teeth on edge. 'I have the honour to report . . . a very extraordinary piece of intelligence', Burnes exclaimed, 'of the arrival in this city yesterday [19 December] of an agent direct from Russia.'[27] Two days later Burnes was able to discover more information on his Russian counterpart. 'The individual who has arrived here from St. Petersburg is a veritable agent of Russia, and brings letters from the Shah of Persia and Count Simonich. He is designated as Captain [Yan] Vikevitch.'[28] This unlikely Russian agent was a Polish-born patriot, who before being sent to Kabul was languishing in Orenburg Prison for having published anti-tsarist poetry. Vikevitch's mentor was the equally improbable German scientist Alexander von Humboldt, who took notice of the young revolutionary's fluency in Persian and several Kirghiz dialects.[29] Von Humboldt persuaded the Russian authorities to free Vikevitch and send him on missions in Central Asia, usually in native dress under the name 'Omar Beg'. Burnes also sent a number of enclosures that were guaranteed to set alarm bells ringing in Calcutta. Through his native spy network, he had intercepted a letter from Dost Mohammed to the tsar, clearly indicating that the emir was prepared to hedge his bets by appealing to Russia to 'arrange matters in the Afghan country'. Likewise, Mohammed Shah had been in touch with the emir, referring to the tsar as his 'esteemed brother' and offering to extend the Afghan ruler the protection of his kingdom. The knock-out blow came in an intercepted letter of introduction from Simonich, presenting Vikevitch to Dost Mohammed, along with a list of 'Russian rarities' to be lavished on the emir that must have raised Macnaghten's parsimonious hackles: a chest filled with finely woven shawls embroidered in gold and silver, something the ladies of the zenana were certain to find more pleasing than Burnes's pins and needles. The reply was not long in coming:

> If he [Vikevitch] be not already gone from Kabul, you will suggest to the Emir that he be dismissed with courtesy. If he [Dost Mohammed] should, on the other hand, seek to retain the agent, and to enter into any description of political intercourse with him, you will give him distinctly to understand that your mission will retire, that out [*sic*] good offices with the Sikhs will

wholly cease and that, indeed, the act will be considered a direct breach of friendship with the British Government.[30]

A chorus of voices now began to bombard the Government with a flow of distressing news. Frederick Mackeson, who bore the peculiar title of Agent for Navigation on the Indus, informed Wade of Vikevitch's plan to visit Ranjit Singh in Lahore for 'friendly conversations', while secretly gathering intelligence on the Sikhs' military power. Robert Leech of the Bombay Engineers, who was with the Burnes embassy to Kabul, found out through intercepted correspondence that the Russians were prepared to finance and equip a joint Persian-Afghan army to attack the Sikhs.

There was yet another shadowy figure who had taken up residence in Kabul at the times of Burnes's mission. The Taliban were not the first to set about defacing the great caves at Bamiyan that housed the giant Buddha statues destroyed in 2001 by Afghanistan's fanatical Islamist regime. In 1930, a French expedition sent to survey the caves was astonished to discover a piece of nineteenth-century English graffiti on the walls:

> If any fool should this high *samooch* [cave] explore,
> Know that Charles Masson has been here before.

Pottinger was sufficiently circumspect to avoid any contact with this dubious character. Burnes, on the other hand, was frequently in touch with Masson, a man very much cut from the same mould as 'Bokhara Burnes'. Masson was an adventurer and traveller, who sometimes went in the wilderness as a starving beggar, on other occasions as a gentleman at ease with all the trappings of wealth. Masson was a pioneering archaeologist and numismatist, whose work enabled later scholars to unfold the history of Buddhism in Afghanistan. He unearthed evidence to prove the existence of Greek kingdoms in the Hindu Kush, legacies of Alexander the Great, and he uncovered a wealth of material to unravel the complex history of the Silk Road and ancient international trade between Europe, India and China. The Masson Collection of several thousand ancient coins is one of the British Museum's numismatic treasures.

Masson was in reality a fugitive on the run, a deserter from the Indian Army. Moreover, he was not even Charles Masson, a US citizen from Kentucky, as he tried to convince Wade when he was detained while passing through Ludhiana. Masson was born James Lewis in Aldermanbury, London, the son of an oil dealer. Wade could have had Masson sent straight to the gallows for desertion, but he saw ways to make use of the intrepid traveller's wealth of contacts and fund of knowledge on Afghanistan. Wade presented Masson with a fait accompli: a full pardon in exchange for an agreement to operate as a spy in Kabul.

We have no record of what was discussed at the bizarre Christmas dinner that Burnes hosted for his enemy Vikevitch in 1837. How one wishes to have been a fly on the wall at that meeting between the espionage agents of the world's two most powerful empires, to have observed Vikevitch and Burnes circling one another like snarling dogs. Nor do we know if Masson was in attendance, though it is unlikely Burnes would have left a European Christian out in the cold that night, if for no other reason than to tip the balance of power in his favour in this triumvirate of spies.

At the outset of their relationship, Burnes wholly endorsed Masson's opinion that Peshawar should be returned to Dost Mohammed, as a prerequisite to stability on British India's North-West Frontier. Masson and Burnes shared a number of character traits: both were opportunists and seasoned survivors. When the Government announced that further negotiations with Dost Mohammed amounted to a waste of time and the only recourse was to declare the mission terminated, Burnes and Masson voiced in unison their view that Shuja could easily get his throne back. Masson went so far as to assure the Government that the Saddozai pretender could be promptly reinstated under British auspices and with little bloodshed. His prediction was correct up to that point, though unlike Burnes he did not stay around to witness the catastrophe that came in the wake of Shuja's restoration. Masson and Burnes departed Kabul together in April 1838, more than a year before the British invasion, which Masson sat out in Sind writing up his research for publication in England. On his return to London in 1842, Masson, then aged 44, married 18-year-old Mary Anne Kilby, a farmer's daughter. On his death in 1853, Masson left behind a collection of some 80,000 rare coins, an unfinished novel, two children and a heartbroken wife, who followed him to the grave two years later at the age of 31.

By March 1838, Burnes's relationship with Dost Mohammed had cooled considerably. In addition to the emir's abiding demand for the return of Peshawar, he now sought British undertakings to protect his kingdom from Persian attack. Burnes knew full well that these terms would be rejected and, moreover, that his usefulness as an interlocutor between Kabul and Calcutta was at an end. He was also keenly aware that his prospects of promotion could in no way benefit from continuing to back the underdog. The way forward was to ingratiate himself with Macnaghten, hence his letter of 13 March assuring the political secretary that 'I would agree to none of the terms proposed', and furthermore, given that he saw 'no hope of adjustment in the present tone held, I should request my dismissal, and proceed to Hindustan'.[31]

The emir was greatly perturbed by Burnes's request to quit Kabul. He assembled all his advisers for a stormy discussion that lasted well past midnight. Dost Mohammed was desperate to rebuild his bridges with the British, who he believed still represented his best hope in the face of Russo-Persian aggression. What he never suspected was that the Government had already settled on a course of action that was to cost the emir his throne. Burnes sensed what was afoot and despite the emir's 'many expressions of regret' (and Burnes's lingering affection for Dost Mohammed), he told his host he was determined to abandon Kabul. This he did on 26 April, under escort of two of the emir's sons.

'Thus ended a mission', wrote Masson, 'one of the most

extraordinary sent forth by a government, whether as to the singular manner in which it was conducted, or as to the results.' Masson goes on to assert that Dost Mohammed 'was most to be pitied'. It was the general opinion in Kabul, and indeed one shared by Masson, that had the emir not been subjected to such ruthless bullying by Macnaghten and his cronies, 'he would have done as much as could be hoped'.[32]

Shortly after Burnes's departure, Dost Mohammed wrote to Auckland to explain the circumstances of the mission's failure. The emir ended his letter with exquisite dignity, stating:

> When Captain Burnes reaches India, he will minutely speak to your Lordship on all the circumstances of this place. There are many individuals who have enjoyed the favour of the British, but our disappointment is to be attributed to our misfortune, and not to the want of the British Government. What is worthy of the good name of the British Government, it, I hope, will come to pass in future.[33]

Burnes and Masson crossed the passes towards Jalalabad, riding through the defiles that were shortly to become the graveyard of a retreating army. At the same time Auckland and Macnaghten, who had by now retired to the pleasant hills of Simla ahead of the summer monsoons, pondered the calamitous state of affairs beyond the Indian borderland. The Persians and their Russian allies stood poised at the gates of Herat. Auckland as well as Dost Mohammed had given up the city for lost, notwithstanding Pottinger's spirited defence. The rulers of Kandahar were moving towards that same city to lend their support to a military venture which in the Government's mind heralded an advance on India. With Burnes's departure from Kabul, the Government had lost all influence at the court of Dost Mohammed. The emir was nursing a grievance against Ranjit Singh over a dispute that could easily erupt into open warfare, a move Dost Mohammed might consider necessary to appease his Barakzai brethren and save his throne.

British India was in urgent need of someone to unravel this Gordian knot. The man to disentangle the quandary with one bold stroke was Lord Palmerston. Britain's foreign secretary was the progenitor of gunboat diplomacy, the ultimate expression of Empire basking in the illusion of invincibility. The most notorious deployment of this policy took place 12 years after the Persian crisis. In 1850, Palmerston dispatched a squadron of warships to blockade the Greek port of Piraeus in retaliation for an outrage committed against David Pacifico, a Portuguese Jew, whose home had been attacked by an anti-Semitic mob. Pacifico had been born in Gibraltar and was therefore a British subject. Palmerston could not allow this affront to go unpunished and within eight weeks the Greek Government had agreed to pay suitable compensation.

Palmerston's use of gunboat diplomacy in the Persian Gulf was on a less picturesque but far more momentous scale. On 21 May 1838, Palmerston instructed McNeill to deliver a final warning to the Shah of Persia, 'to declare to him explicitly that the British Government cannot view with indifference his project of conquering Afghanistan'.[34] This warning was reinforced with a follow-up letter stating in no uncertain terms that the shah's enterprise constituted an act of hostility against the British Crown. The inevitable outcome was one month in coming: Palmerston sent the gunboats with a detachment of troops to occupy the island of Karrak in the Persian Gulf, where the soldiers received an effusive welcome from the island's horrified governor.

For good measure, Palmerston fired off a letter to his Russian counterpart Count Karl Robert Nesselrode, reminding him that Great Britain always regarded Persia as a barrier for the security of British India, and that Russia and England had an agreement to consult one another on Persian affairs. Palmerston then pointed out that Count Simonich and Vikevitch were acting in a manner contrary to good relations between both imperial powers. Palmerston told Nesselrode that he was fully aware of what both these agents were up to and that Britain was 'too conscious of her own strength and too sensible of the extent and sufficiency of the means which she possesses to defend her own interests in every quarter of the globe' to be intimidated by Russian meddling in Persia. He unblushingly demanded to know whether Russian policy was 'to be deduced from the declarations of Count Nesselrode ... or from the acts of Count Simonich and M. Vikevitch'.[35] Nesselrode must have been beside himself with rage to receive such an audacious piece of chutzpah. But apart from a timid retort, in which he accused Britain of spreading 'disquiet' in Central Asia and trying to deprive Russia of markets, Nesselrode left no doubt in his reply that Russia lacked any appetite for open confrontation with England. In compliance with a strongly-worded note from Britain's Ambassador to St Petersburg, the Marquess of Clarincarde, Nesselrode recalled Simonich, who swiftly vanished into obscurity in the cavernous halls of the Foreign Ministry. Vikevitch was summarily disowned by Nesselrode, who even refused to receive him on his return to St Petersburg, whereupon the disgraced Russian agent returned to his hotel and put a bullet through his head with his service revolver.

As for Mohammed Shah and his pretensions of territorial conquest in Afghanistan, the Royal Navy's token occupation of a piece of Persian territory had the desired effect. The siege was given up when the shah realized that not only was he on a war footing with Britain, but also that he could no longer count on Russian support. The news was delivered by Lieutenant Colonel Stoddart who, before his tragic murder in Bokhara four years later, was in 1838 on service with the British diplomatic embassy to the shah's court. On 9 September of that year, from his position close to the shah's encampment, Stoddart reported to McNeill that the Persian Army had lifted the siege of Herat. His message ended with the laconic postscript: 'The Shah has mounted his horse *Ameerij* and is gone.'[36]

CHAPTER 2

Victoria's First War

Lord Auckland was not a man to be distracted by technicalities when it came to pursuing a belligerent policy towards Afghanistan. The fact that the Shah of Persia had abandoned the siege of Herat and also that his Russian comrades-in-arms had withdrawn all diplomatic and military support for Persia's ambitions did not deter the governor general from making preparations for war against an imaginary enemy. In fairness, Auckland was spurred on by a tight little group of hawks, led by Palmerston, McNeill, Macnaghten and Sir John Hobhouse, president of the East India Company's all-powerful Board of Control.[1] Palmerston openly avowed marching an army straightaway into Afghanistan. In a note to Hobhouse after learning of Mohammed Shah's abrupt retreat from Herat, the foreign secretary stated: 'It is evident that Afghanistan must be ours or Russia's, and this time we have the choice in our own hands.'[2] Hobhouse, through the Board of Control's Secret Committee, unequivocally instructed Auckland to oust Dost Mohammed and restore Shah Shuja to the throne of Kabul. Hobhouse cloaked his declaration of war in a gesture of compromise, by asking Auckland to consider making a final attempt at reconciliation with Dost Mohammed, though the Government was now fully convinced the emir was conspiring with the Russians. This offer, Hobhouse stated, should be presented to Dost Mohammed in the form of 'final demands'. He then added that 'If these were rejected, the Army should cross the frontier.'[3] It would be difficult to see how Auckland could have opposed this chorus of Britain's most belligerent political leaders, all braying in unison for war.

The die was cast and it was decided that Dost Mohammed was to be removed as a 'threat' to British India's security. In his place, the Saddozai dynasty would be restored in the person of the superannuated Shuja. The British chose to ignore the fact that Shuja was in every respect a loser. Less than a decade had passed since Shuja's last inglorious attempt to recover his throne had ended in disaster. That military expedition was officially disowned by Calcutta and the authorities at Ludhiana, where the former emir was luxuriating in gilded retirement with his several hundred wives. In the end, despite official assurances that 'the British Government religiously abstains from intermeddling with the affairs of its neighbours',[4] Macnaghten agreed to give Shuja a four-month advance on his stipend to enable him to raise a body of troops from his Saddozai tribal following.[5] Shuja negotiated a treaty with Ranjit Singh, pledging the king of the Sikhs his unconditional allegiance and stipulating that each would show mutual respect for the territories under their respective control. Ranjit Singh drafted a treaty allying the two powers against Dost Mohammed and promising to treat the enemies of either one as a common foe.[6] Shuja put together a force of some 30,000 men, who set out on a campaign across Sind to invest Kandahar, where they were soundly trounced by Dost Mohammed's horse and infantry, leaving 5,000 dead before the city walls. By day's end the Saddozai troops, 'a feeble, broken-spirited rabble',[7] abandoned the field, and Shuja turned his elephant around and marched back to Ludhiana. For his part, Ranjit Singh had agreed to relinquish all claims to the lands held by Shuja, but as this did not include Peshawar the maharaja lost no time in incorporating that city into his kingdom.

On 28 June 1838 Queen Victoria was crowned in Westminster Abbey. Five months later commenced the first war of her long reign. Though only a girl of nineteen, Victoria was sagacious enough to sense that this part of the world was destined to feature large in the history of her kingdom and hence in her own life. A month after the declaration of war, she requested the Foreign Office to send her a map of Asia.

Kaye aptly described Simla as 'that pleasant hill sanatorium where our Governors General, surrounded by irresponsible advisers, settle the destinies of empires . . . and which has been the cradle of more political insanity than any place within the limits of Hindustan'.[8] If one were searching for a single document to illustrate this claim, one would need to look no further than the Simla Manifesto, issued on 1 October 1838. Kaye's was far from being a lone voice in the wilderness. The Simla Manifesto, which bore Macnaghten's signature (and blessing) as political secretary, has variously been described by historians and soldiers past and present as 'a travesty of the facts', 'unscrupulous', a 'discreditable enterprise' and 'an act of obsession'. No lesser an authority than Mountstuart Elphinstone considered it 'hopeless' to try to maintain Shuja on the throne of Kabul. The Duke of Wellington prophesized that crossing the Indus to set up a puppet government in Afghanistan would precipitate a perennial march into that country. He was not the only military leader to condemn the Manifesto and the proposed invasion. 'Never before, during the history of the British power in India, had so wild, ill-considered, and adventurous a scheme of far-distant aggression been entertained', thundered Major General Sir Henry Mortimer Durand, who was attached to the invasion force that marched under the resplendent title of the Army of the Indus.[9]

In this pronouncement, Auckland asserts that he intended to use his good offices with Ranjit Singh 'for the restoration of an amicable understanding' with Dost Mohammed, while in fact the Government of India never made the slightest effort to advance the Afghan case. As for the Persian attack on Herat, Auckland maintains the Afghans were conniving to extend Persian influence 'to the banks of, and even beyond, the Indus', but he fails to offer a shred of evidence to substantiate this claim. The governor general said that Dost Mohammed 'avowed schemes of aggrandisement and ambition injurious to India' and that the emir threatened to call in 'every foreign aid which he could command', while ignoring the fact that this 'foreign aid' (i.e. the Russians) had vanished from the scene. Auckland brought his rant to an end by accusing the Barakzai clan of being 'ill fitted to be useful allies to the British Government'. This is, at the least, a monumental travesty. Dost Mohammed, apart from two years' forced exile in India, occupied the throne of Kabul for 37 years, the longest reign in Afghan history, and he is today revered as one of the country's national heroes. Almost every allegation put forward in this spurious document is half-baked malarkey, when not an outright fabrication.

The most outrageous mockery of the truth contained in the Simla Manifesto was Auckland's charge that Burnes had been compelled to abandon his mission to Kabul as a result of Dost Mohammed's 'utter disregard of the views and interests of the British Government'. It may be argued that Burnes's mission to Kabul had been foredoomed to failure. Burnes had established the most cordial relations with the emir, but there his success had stopped. 'Between him and Lord Auckland there was a hopeless incompatibility of purpose', notes one historian. 'To Auckland, the British alliance with Ranjit Singh . . . was, diplomatically speaking, the ark of the covenant . . . Burnes wanted the British to strengthen Dost Mohammed's local authority, while Ranjit Singh's interests were best served by Dost Mohammed's weakness.'[10]

The Simla Manifesto, which purported to bestow credibility on the Afghan adventure of 1838, has been denounced as irresponsible, deceitful and unjust, but it could also be condemned as a criminal act. Years later, when he came under pressure from a group of MPs demanding an inquiry into the invasion, Hobhouse gave evidence before the House of Commons. He stood up and proclaimed himself responsible for the war, meaning he had personally incited the Government at home and in India to march an army into Afghanistan to depose Dost Mohammed. The reason given by the president of the Board of Control was 'the inveterate hostility' of Dost Mohammed to the British. This was simply not true. What had transpired in the run-up to Lord Auckland's manifesto was this: Burnes's reports were edited and effectively falsified, deleting references to the emir's desire for friendship with Britain. Burnes's last letter to Macnaghten spoke of the Amir as a 'man of ability', with a high opinion of the British nation. Burnes forthrightly stated that if the Government were to do half for Dost Mohammed what they did for Ranjit Singh, the Afghan emir 'would abandon Russia and Persia tomorrow'. This letter was omitted altogether from the Government Blue Book published before the invasion. It later emerged that Hobhouse himself was the culprit who had doctored the letters, making it appear that Burnes had reported hostility on the part of Dost Mohammed and had recommended going to war. Burnes became aware of the falsification in late 1839, when he sent the true copies to his brother for safekeeping. In an accompanying letter, Burnes said that the governor general's views as they appeared in the Parliamentary Papers of the day were 'pure trickery'. But the fraud was not exposed to the public until 1861. In that year the outspoken Scottish MP Alexander Colquhoun-Stirling-Murray-Dunlop requested Lord Palmerston to open a formal inquiry into the forgeries. He was shouted down by the Establishment, from *The Times*, which said that 20 years after the event the Government should consider this a closed matter, to Benjamin Disraeli, who also argued that the subject was unworthy of debate. In 1861, Burnes's brother James fired off a series of angry letters to Palmerston, who was serving his second term of office as prime minister. James Burnes sought 'to disprove the unfounded assertions made by him [Palmerston] not four months ago, with the view of covering the dishonesty then exposed'. He demanded that Palmerston explain his remarks made to the House, which he interpreted as a smear on his brother's character and, by implication, the truthfulness of his dealings with Dost Mohammed. After an exchange of eight letters came the final reply from Downing Street. It stated, 'Lord Palmerston has no wish to enter into further discussion on this subject.'[11]

As a footnote to this sorry episode in British diplomacy, the Simla Manifesto evokes a number of uncomfortable parallels with an event much closer to us in time, one which concerns an attack on another country in the same part of the world. This was a forerunner of the 'dodgy dossier' utilized to justify Britain's participation in the 2003 invasion of Iraq, with the 'Russian threat' serving as the fanciful 'weapons of mass destruction' and the identical objective of achieving 'regime change'. *Plus ça change*, one would rightly be inclined to think.

Five months before the Simla Manifesto was conceived, Auckland put together what in current diplomatic parlance would be termed a 'coalition force', in reality an alliance of British and Indian Army divisions supported by token contingents of Afghan and Sikh troops. Auckland dispatched Macnaghten to negotiate a treaty between the British Government, Ranjit Singh and Shuja. Macnaghten journeyed to Lahore to explain the governor general's scheme to the Sikh ruler, who received him with great pomp and ceremony. Ranjit Singh immediately took to the proposal of making Britain a party to the earlier treaty between himself and Shuja. 'This would be like adding sugar to milk', the old greybeard exclaimed.[12] Macnaghten went on sweetening the milk, with assurances that the governor general was prepared to help his ally Shuja raise an army of loyal Saddozai soldiers and supply officers and money for the undertaking. The strategy called for Shuja's column to march on Kandahar, where he had so miserably failed in his previous endeavour to regain his crown. The Sikh Army was to move on Kabul via Peshawar and the Khyber Pass. Here Ranjit Singh demurred: if the invasion was successful, would this not result in increased British influence in the region, as well as greater power for Shuja, to the possible detriment of the Sikhs? For the moment, the maharaja gave Macnaghten his reluctant consent. In the end, Ranjit Singh managed to extract a key concession from the British, an agreement not to send their army through the Punjab.

With Ranjit Singh's signature barely dry on the treaty, Macnaghten galloped across the plains in the monsoon downpour, 100 miles south-east to Ludhiana. In his encounter with Shuja, he regaled the despondent royal exile with the same lies he had announced to the maharaja, namely that Burnes had been cold-shouldered by Dost Mohammed, who had rejected Burnes's offer to mediate in the emir's conflict with the Sikhs. Macnaghten insisted this hostile attitude necessitated installing a friendly ruler in Afghanistan.

Shuja required no encouragement to jump on the bandwagon – after all, this time he was being escorted free-of-charge back to Kabul under the protection of the British Government. Macnaghten described the need to make the Sikhs a party to the venture, given Ranjit Singh's expanded sovereignty over much of the old Afghan Empire. The political secretary spent an anxious moment listening to Shuja enthuse over the new treaty, for in the same breath Shuja reasoned that once he was restored to power, it would be logical to expect the Sikhs to hand over Peshawar to its rightful owners, the Afghans. The crafty Macnaghten eventually persuaded Shuja to yield on this demand, by enlisting Shuja's aides to tactfully point out that this would contravene the existing treaty with Ranjit Singh. Then to drive home the message of just who was the junior partner in this joint venture, a clause was inserted whereby Shuja agreed to show his gratitude for Sikh support by providing the maharaja with an abundance of gifts each year, along with a cash payment of 200,000 rupees to support 'a Sikh force of not less than 5,000 men . . . of the Mohammedan persuasion within the limits of Peshawar territory for the support of the Shah' and to safeguard his eastern frontier.[13]

The gift list called for the yearly delivery of 55 high-bred horses 'of approved colour and pleasant paces', 11 Persian scimitars, seven Persian poniards, 25 good mules, fruits and nuts of various kinds, as well as satin, furs, carpets and other items to the total of 101 pieces.

In late November 1838, the Army of the Indus gathered en masse at Ferozepore, a town in the Punjab on the banks of the Sutlej, for a splendid send-off to Afghanistan. The crowning moment was to be an historical face-to-face meeting between Lord Auckland and Ranjit Singh, the leaders of two great empires, standing together against a common enemy. General Sir Henry Fane, Commander-in-Chief, India, a fine soldier of the old

Tory school, rode into camp two days ahead of the great event to oversee the final preparations. He arrived amidst a scene of uproar and confusion, the deafening clang of horns and drums, and the trumpeting of two rows of elephants lining the road to the Durbar tent.[14]

Presently, Ranjit Singh made his appearance on a caparisoned elephant, surrounded by a retinue of secretaries and military commanders. It was to be a grand moment when Auckland came forth to greet his Sikh ally and the two elephants were drawn side by side. Instead, what ensued was a scene of mayhem that came perilously close to a disastrous climax, as the British and Sikh cortege of elephants were mixed up together in one great crush. The Sikh bodyguards feared an assassination attempt on their ruler and many reached for their weapons. The sight of Ranjit Singh, who had climbed down from his beast, restored calm to the occasion. Auckland then presented the maharaja with two nine-pounder guns. The diminutive Sikh promptly stumbled and fell in front of the muzzles of the British guns. This set the guards into another frenzy and it was only Fane's quick diplomatic intervention that prevented the ceremony ending in a Wild West-style shootout. This is how Kaye sums up the end of the first day's proceedings: 'Then, amidst an uproar of hurrahs, a crash of military music and another scene of indescribable confusion, Ranjit Singh ascended his elephant and turned his back upon the British camp.'[15]

The following day it was Auckland's turn to pay a visit on his ally's great ceremonial tent, where the governor general and his sister Emily were treated to a spectacle of dancing girls and 'low vices', which the British contingent observed with a blush. Emily Eden described the nautch girls as 'a very ugly set from Ludhiana' and Ranjit Singh as 'an old mouse, with grey whiskers and one eye'. Auckland's sister was seated next to Ranjit and had a very glum time of it. 'I could not talk with any great ease', she complained, 'being on the blind side of Ranjit, who converses chiefly with his one eye and a few signs which his faqir makes up into a long speech.' She was even less impressed by the man seated on her other side, who was destined to become Ranjit's successor. 'Kharak Singh was apparently an idiot.'[16]

The Army of the Indus was the largest invasion force ever assembled under the Raj, numbering upwards of 39,000 men of the Bengal and Bombay Armies. This was before taking into account a tail of 38,000 camp followers, with thousands more of pack animals, strung out for miles behind the columns of troops. Some elite British regiments marched with the main body, including the 16th Lancers, complete with their pack of fox hounds. At Ferozepore the expedition became known as the Grand Military Promenade, so confident was the soldiery of embarking on a walkover campaign. But a mood of dejection set in amongst the ranks even before the march got underway, for it was learnt that several thousand of their colleagues were to be left behind when the army set out to Afghanistan. On 27 November, Fane announced that some regiments would be held back due to the 'altered state of affairs' west of the Indus. The 'change' the commander-in-chief referred to was the lifting of the siege of Herat and the retirement of all Persian and Russian troops from the 'gateway to British India'. Auckland later explained in testimony to the Secret Committee that in view of this unexpected development, he had decided to retain part of the Bengal Army at Ferozepore 'and to direct the immediate movement on Shikarpore [a town in Sind Province] of only the first division under the command of General Sir Willoughby Cotton' with some 7,500 men.[17] It had apparently never crossed Auckland's thoughts that the entire operation had lost its *raison d'être*. No, war would be waged regardless, for it was necessary to secure India against foreign aggression – but from whom? The Russians and Persians were gone, and Dost Mohammed lacked the resources to take on his Sikh foes, much less an army of disciplined, battle-hardened British and Indian troops.

Ranjit Singh returned to Lahore in the company of Lord Auckland, who had accepted the maharaja's invitation for a state visit to the Sikh capital. Before quitting Ferozepore, Auckland appointed Macnaghten British Envoy to the court of Shah

Shuja in Kabul, with Alexander Burnes as second-in-command. Burnes must have regarded this as a slap in the face, after having abandoned his friend Dost Mohammed, the man he believed in, to throw in his lot with the Government.

Cotton took the advance, following the course of the Indus to rendezvous with General Sir John Keane, who was in command of the Bombay force. Shuja's levies, consisting of 6,000 cavalry, infantry and horse artillery, rode with Cotton on the trek to the Bolan Pass, a journey that took the army some 600 miles in a clockwise direction away from its initial objective of Kandahar. This was the first of a catalogue of errors – apart, of course, from the decision to launch the invasion – that rendered the Afghan venture one of Britain's costliest military disasters. Political considerations had dictated the choice of taking this circuitous route, instead of moving due west across the Frontier passes to Kandahar, for the Government saw this as an opportunity to force the recalcitrant emirs of Sind into submitting to Shuja's authority.

Fane was by now losing enthusiasm for the whole undertaking. From a purely tactical standpoint, the army was heading into a vast desert inhabited by hostile tribes, which the troops would have to cross at a smart pace to avoid getting mired under their own weight. Despite Fane's pleas, even junior officers insisted on taking with them as many as 40 servants each and as many camels per man.

The first British casualty of the Afghan campaign was the army's commander. At 60, Sir Henry Fane was in those Victorian times viewed as a man on the threshold of old age. Yet he was far from being the army's only sexagenarian on active military service. Sir Charles Napier achieved his famous victory in Sind at the age of 61 and at 70 went on to become Commander-in-Chief, India. Fane's successor, Sir Jasper Nicholls, was appointed at the tender age of 61. Fane was in failing health and for some time he had been thinking about retirement. But there were other more compelling reasons for him to relinquish command of the Army of the Indus. He began to doubt the rationale for an invasion as soon as Herat had been put out of danger. On the other hand, once a massive force had been assembled to do the job, Fane saw no sense in placing the troops at greater risk by paring back on numbers. For the top-ranking military man in India, Fane was remarkably restrained in his views on imperial expansion. He once famously stated, 'Every advance you make beyond the Sutlej to the westward, in my opinion, adds to your military weakness . . . Make yourselves complete sovereigns of all within your bounds. But let alone the far west.'[18] Fane's hostility towards Macnaghten exposed the first cracks in the military-political rift that was to prove calamitous to the Afghan expedition. Fane told Auckland he didn't think his 'service is needed' for the governor general's proposals. He also had grave misgivings about the authority conferred on Macnaghten: 'I think too that your instructions to Sir William Macnaghten and to me are such that an officer of my rank could hardly submit to serve under.'[19] Auckland was unruffled by his military commander's censure of Government policy. Quite to the contrary, in a report to the Secret Committee on the appointment of the Bengal Army's Keane as overall commander of the Army of the Indus, Auckland twists Fane's resignation into 'the services of his Excellency General Sir Henry Fane having been dispensed with . . .'.[20] Fane now felt himself too weak to carry on in India's insalubrious climate, so he sailed from Bombay on home leave in early 1839, and died at sea off the Azores.

The Government had some years previous signed an agreement with the emirs of Sind, by which the British were given access to the Indus for commercial traffic, while expressly forbidding military movements on the river. This treaty was cast aside as Keane landed his Bombay division at Karachi in late November, to cross the Indus and link up with Cotton's Bengal troops in the Bolan Pass. The hero of Herat, Eldred Pottinger, was toiling feverishly to smooth the feathers of the emirs over this treaty violation while Burnes, whose timely decision to tow the Government line had earned him a knighthood and a promotion to lieutenant colonel, deployed his considerable diplomatic skills on the equally annoyed tribal chieftains of Baluchistan.

Shah Shuja and his contingent crossed the Indus in early 1839, where he was joined at Shikapur by Macnaghten. The military authorities now determined that the bulk of the Bengal column should proceed down the left bank of the Indus to co-operate with Keane to subdue the rebellious post of Hyderabad, a campaign that was cancelled when the local emirs submitted to British terms. But Macnaghten did not approve of the diversion and he took upon himself the responsibility of preventing the Bengal Army's march along this route. 'The two columns were entirely ignorant of each other's operations in the absence of an Intelligence Department, a want which continued to be felt until the close of the war', according to an army report on the campaign. 'The military and political authorities were brought into a state of undisguised antagonism.'[21]

The Bombay and Bengal contingents finally joined forces at a point about ten miles up the Bolan Pass on 4 April, both columns having suffered severely from a lack of provisions. Cotton had placed his troops on half rations, and half again for the camp followers. By the time he caught up with the Bombay Army and its commander Sir John Keane, Cotton had reduced his entire force to quarter rations. Moreover, two-thirds of the army's 30,000 baggage camels had been lost to thirst, hunger, overwork and theft by raiding Baluch tribesmen. The lack of water was an acute problem for the troops on their trek across the deserts of Upper Sind. Some of the soldiers were driven to near insanity, to the point that on one march, several were found dead under a tree. They had tried to quench their craving by drinking a bottle of brandy in the blazing heat.

It is truly a wonder that Cotton and Keane were able to campaign together in harmony. Keane was a man of violent temper, who addressed his colleagues in the language of the barrack room and had a deep contempt for native Indian soldiers. His uncouth manner earned him the enmity of Auckland, and for this reason he was passed over for appointment as Commander-in-Chief, India. That said, Keane was the only senior officer with the audacity to prophesy a disastrous outcome for the Afghan campaign. Cotton was of an equally pugnacious, though less adroit nature. Macnaghten disliked him intensely and called him a 'sad croaker'. The envoy took umbrage at Cotton's assurances that Shuja was unpopular in Afghanistan and that the army would meet stiff opposition from Dost Mohammed's supporters. The general was to be proved mostly correct in his predictions. Fane as well found Cotton a bit on the dim side and cast doubt on his ability to execute verbal orders. Cotton lacked Keane's personal charisma and was in fact regarded as a tiresome character, forever retelling in the mess his favourite anecdote of how he was expelled from Rugby School for inciting his schoolmates to burn the headmaster's desk and books. So together they marched, the club bore and the brute, at the head of a half-starved army, across the baking Kandahar plain.

Averaging ten miles per day, it took the army six days of constant climbing to negotiate the Bolan Pass. The advance of the Bengal column reached Quetta in the first week of April. The Bombay contingent had an even less pleasant crossing, the trail being strewn with the decomposing bodies of camels and horses, as well as having to suffer a shortage of supplies of every kind. Keane had no wish to linger at Quetta, where food and water were still in short supply. The commander was for pushing on to Kandahar without delay. Macnaghten was in full agreement on moving quickly against the Barakzai stronghold. Pompous and possessed of an insane vanity he was indeed, but the envoy was not a stupid man. He recognized the importance of having Shuja securely installed on the throne, the sooner the better. Macnaghten had had ample time on the march from India to observe the Saddozai pretender at close quarters. He was now beginning to acknowledge the wisdom in Cotton's assessment of the Government's protégé. On Shuja, Macnaghten wrote, 'His opinion of the Afghans as a nation is, I regret to say, extremely low. He declares that they are a pack of dogs, one and all, and as for the Barakzais, it is utterly impossible that he can ever place the slightest confidence in anyone of that accused race.' As for the need to get on with the campaign, Macnaghten stated, 'The fact is, the troops and followers are nearly in a state of mutiny for food.'[22]

On 7 April 1839, Keane put his army back on the move. The columns laboriously ascended the 7,500-foot Kojuck Pass that rises between Quetta and the town of Chaman. Shuja, bubbling with anticipation, rode in the advance. Morale in the ranks was greatly lifted as supplies became more plentiful – the local tribesmen were delighted to attend to the appetites of a mass of hungry men. Uninvited infidels for sure, but far be it from an Afghan to turn his back on a sale at inflated prices. Within one week the army stood poised on a ridge above the Kandahar plateau, their objective almost in sight. After a well-deserved rest to make ready for the attack, by 23 April the force was at last before the city, with spirits running high at the prospect of a good fight. For the past five months the troops had suffered hunger, thirst, illness and boredom, with hardly a shot fired at an enemy, apart from repelling the odd band of marauders. Now at last they were to enter into combat – except they weren't.

Much to Macnaghten's elation, the military commanders' dismay and the astonishment of all, the city was found to be totally undefended. Kandahar is the sacred abode of the Pashtuns, Afghanistan's ruling tribe since time immemorial. The citadel has always been the prize of Asia's greatest conquerors, from its founder Alexander the Great in the fourth century BC to Genghis Khan and Mahmud. In the eighteenth century, it became the capital of a sprawling Afghan empire that stretched across Asia from Persia to Kashmir, and more recently it was the spiritual homeland of the Taliban. But in 1839, Kandahar's sirdars, with Dost Mohammed's brother Kohun-dil-Khan in the vanguard, had fled the city without putting up resistance, to seek sanctuary with their old ally Mohammed Shah in Persia.

The army marched in great triumph through the gates in a scene reminiscent of Mountstuart Elphinstone's entry into Peshawar exactly 30 years before. Shuja could hardly contain his joy on his procession through the streets, accompanied by Macnaghten and the principal army officers. There was a vast assemblage of gazers, straining to catch a glimpse of the returning monarch. The women clustered in the balconies of the houses or gathered on the rooftops, while the men thronged the streets below. The people strewed flowers before the horses and loaves of bread were scattered to the soldiers. Shuja rode up to the tomb of his grandfather, Ahmed Shah, the founder of the Afghan Empire, to offer up thanksgivings and prayers. It was a moment of great relief for Macnaghten for here, before his eyes, he had witnessed the vindication of the Government's much criticized determination to take Britain to war. 'We have, I think, been most fortunate in every way', the Envoy communicated in a dispatch to Auckland in Simla. 'The Shah made a grand public entry in the city this morning and was received with feelings nearly amounting to adoration.'[23] Macnaghten found himself obsessed with the preparations for this glorious day, to the point that he had been unable to sit down to breakfast before three in the afternoon.

A great ceremony was organized for 8 May to celebrate the restoration of the Saddozai dynasty. Both columns of the army were turned out before Kandahar's walls. After taking the royal salute, Shah Shuja solemnly ascended a raised platform to review the troops, against the horrific din of 101 guns that thundered across the plain. When the British regiments were dismissed, the drums ceased to roll and the dust kicked up by the horses had settled to earth, Macnaghten found himself staring at a disturbing truth: his entire meticulously planned exercise had been a painful failure. Only a handful of the Afghan citizenry had come out to witness the spectacle. The apparent general enthusiasm that had greeted Shuja's return had now dissipated into sullen indifference. Captain Henry Havelock, a devout evangelist and Cotton's aide-de-camp, was heard to remark that the outburst of fervour at Shuja's presence at the ceremony was in reality confined to his retainers, noting that the space reserved for the populace behind the throne had remained empty throughout the morning's celebrations.

How could it have been otherwise? Had Shuja come to claim

his throne at the head of a victorious Afghan army, he might have stood a chance of winning over his reluctant subjects on the strength of his dynastic credentials. Whatever his personal shortcomings, he was a direct descendant of Afghanistan's national hero. After all, the people of Kandahar were not thrilled to be living under a clique of tyrannical sirdars. But to bring in tow a force of tens of thousands of Feringhees,[24] to whom Shuja owed his allegiance, was taken as an insult to Afghan pride and the Muslim faith. With unfortunate though forceful (and characteristically pompous) hindsight, *The Times* saw through the folly of foisting a British ward upon a people who cherished their pride above all other things:

> The choice of Lord Auckland and his employers lay between the alliance of this able, powerful, popular and established Sovereign with his people [Dost Mohammed], and a military occupation of Afghanistan, for the purpose forcing upon the unwilling and indignant inhabitants a mere creature of the British Government, to be maintained upon his throne by British arms and controlled by British dictation.[25]

Up to now, Keane owed his success more to luck than any tactical superiority over his Afghan foes. Provisions had been found in time to restore his famine-stricken army to health and a good fighting spirit, but more importantly, he had encountered the enemy in a state of complete disarray. Internecine strife had prevented the Barakzai clans from putting up a united defence against the infidels and their traitorous lackey Shuja. Dost Mohammed had little faith in his Kandahar brethren, while the sirdars contemplated the King of Kabul with deep suspicion. For this reason they chose to make their escape to Persia rather than taking their followers to Kabul to join forces with the emir, who for his own part was beleaguered by rebellion in Kohistan to the north and amongst the Kuzzilbash[26] clan at home.

Everything seemed to be going in the army's favour, but then in late June, the day before Keane's advance on the great citadel of Ghazni, there occurred an event that effectively split one leg from the Government's tripartite alliance.

Ranjit Singh had suffered a bout of seizures in the preceding weeks. Neither prayers nor medical treatment were proving of any help to the ailing maharaja. His astrologers advised him to distribute charitable gifts, while his physicians covered his body with agate, all to no avail. In his final dying act, the old, grey-bearded warrior summoned his courtiers to his bedside and instructed them to give away all his weapons, before purifying himself with Ganges water. On 26 June, the 'Lion of the Punjab' slipped into unconsciousness and the following evening he breathed his last, exactly on the 40th anniversary of his triumphant march into Lahore. The loss of Britain's long-standing Sikh ally raised concerns for the security of Afghanistan's eastern frontier. It also opened the floodgates to turmoil in the Sikh dominions, where the line of succession was not clear and the army was rife with factionalism. Two main factions, the Dogra[27] clan and the Sikh aristocracy, fought for the crown. Ranjit's son Kharak Singh, whom it will be recalled was regarded by Emily Eden as an 'idiot', was chosen as the first successor. It took only two months for Kharak's son, Nau Nihal Singh, to ease his opium-addicted father off the throne. Kharak died a day later, but he might have taken heart to know that his son enjoyed a very short reign. Within 24 hours Nau Nihal was also dead, crushed by falling masonry. The throne then passed to Sher Singh, an impressive figure who once sat for a portrait by Emily Eden.

Keane received the report of Ranjit Singh's death on 1 August 1839. By that time, however, other developments had taken place to compensate for the loss of this ally and its potential implications for the war effort. Keane knew that taking Ghazni, the army's next objective, could not in any manner be defined as a military promenade. Ghazni was regarded as Asia's most impregnable fortress, a description curiously at odds with the intelligence given by Macnaghten's 'politicals', who assured the General that the citadel's defences were 'despicable'. Scouts had brought in reports of a body of around 6,000 Afghan horsemen and infantry massed at the fort. The numbers favoured the attackers, however Keane was bringing up a travel-weary army to confront a well-fed and rested foe waiting to give battle with the fanaticism of soldiers fighting for their homeland.

The Macnaghten clique persuaded Keane that whatever he may have heard about Ghazni's formidable reputation, the city could easily be taken by assault. This was the basis for the commander-in-chief's decision to leave behind his siege guns, thereby speeding up the march from Kandahar. Keane struck camp on 27 June, advancing north from Kandahar 230 miles along easy open ground. A little more than three weeks later the city's famed minarets loomed into sight. Within 24 hours of their arrival, the troops found themselves engaged in a brisk exchange of shots with the determined Afghans firing their long-barrelled *jezails* from the battlements. It quickly became apparent that nothing short of a proper siege train would be enough to bring down the walls. In fact, the four guns Keane had left behind in the ordinance depot at Kandahar would have done nicely.

Keane, Cotton and a small party of officers toured the defences one morning in search of some chink in the citadel's armour. They returned discouraged by what appeared to be an unassailable fortress capable of resisting even a protracted bombardment from the army's small field guns. There wasn't much time: the British had no idea how much food and water the Afghans had stockpiled in the fort, but it was clear to all that the army could not afford to remain immobile with only three days' supplies left in the commissariat. There was also the constant danger of Dost Mohammed making an appearance with his sons and Barakzai warriors.

Keane was still on his winning streak. By luck, a disaffected nephew of Dost Mohammed, one Abdul Rashid Khan, made his way to the general's tent with news of that elusive weak point in Ghazni's defences. The Kabul Gate, he informed Keane, had not been properly fortified, since it was here that the Afghans expected to receive Dost Mohammed's reinforcements. The gate was merely barricaded. Blow it open, he declared to the captivated general, and Ghazni is yours. There was no time to lose. Keane issued orders for the assault to begin before daybreak on the following morning, with the outspoken Captain Henry Durand, the officer who had poured scorn on the entire Afghan adventure, leading a company of sappers and miners.

In those same hours Shah Shuja carried out an act of barbarism that left many in the British camp wondering if they had perhaps backed the wrong horse. A war party of the fearsome Ghazis[28] formed up within sight of the tents, ready to sweep on the traitorous Shuja. They were beaten back by two British brigades, the cavalry of Captain Peter Nicholson and foot soldiers commanded by General James Outram. Some 50 prisoners were brought into camp. Shuja reviewed the defiant Ghazis and to the horrified eyes of the British offers in attendance, he had the lot hacked to death by his executioners. Kaye points out that while Macnaghten had been commending the king's humane instincts, 'The humanity of Shah Shuja was nowhere to be found except in Macnaghten's letters.'[29]

A sharp wind howled across the plain that night as Keane positioned his light field guns on a hillock overlooking the fortress. The troops made their way towards the walls under cover of darkness and the roaring wind. The main column advanced under Brigadier Sir Robert Sale, one of the outstanding heroes of the Afghan campaign, who on this day was to receive the first of three wounds before the war was over. By three in the morning, all was in place for the attack. Keane opened with a diversionary barrage on the walls while Durand and the British engineers piled up their powder bags at the Kabul Gate. There was a moment of panic when the powder failed to ignite. Durand boldly ran back to the wall under fire to touch off the fuse with his fingernails.

Down with a deafening crash came tons of masonry and beams. Amid the screams of the wounded Afghans slowly burning to death under the debris, and the cheers of the soldiers poised with fixed bayonets at the gaping hole, the order was given for the bugler to sound the charge. But the unfortunate subaltern had just been shot through the head. The indomitable Durand once more rose to the occasion by finding a replacement, upon which the troops rushed forward. There was one moment of heart-stopping confusion when the passage was found to be blocked with the wreckage of the gate, and this gave the Afghans valuable time to regroup. Sale was the classic general who led from the front and this cost him a sabre cut to the face. After two hours of desperate hand-to-hand fighting, the British took the day. Keane and Macnaghten serenely rode through what remained of the Kabul Gate to raise the Union Flag above the fort shortly before daybreak. Keane reported his success to Auckland in rapturous prose, acquainting the governor general with 'one of the most brilliant acts it has ever been my lot to witness during my service for forty-five years, in the four corners of the globe, in the capture by storm of the strong and important fortress and citadel of Ghazni', highlighting moreover that this feat had been accomplished against all the odds by 'British valour, in less than two hours'.[30] Keane makes mention of the Ghazi attack on Shuja's contingent, but not wishing to put a wet blanket on the occasion, he has deleted any reference to the emir's cold-blooded butchery of the Ghazi prisoners.

It was unthinkable, but Ghazni had fallen to an army of infidels, and now it only remained to feed and rest the troops in anticipation of the coming of Dost Mohammed's host. As it turned out, the emir had altered his plans. Keane learnt on 1 August that the Barakzai force from Kabul was not galloping south after all. There was no longer any need to forcibly drive Dost Mohammed from Kabul, a task that would have cost both sides a great deal of bloodshed. The emir had been left desolated by the fall of Ghazni, yet he made one final attempt to rally his band of fighters, exhorting them to expel the Feringhees or die in the glorious attempt. There were few takers for such a suicide mission.[31] Dost Mohammed, who had enjoyed the adoration of his people for 13 years, now stood alone. There could be only one course of action – the emir released his men from their oath of allegiance, asking only those who remained faithful to him and to Islam to follow their leader into exile. With that, Dost Mohammed and his band of followers turned their horses northwards to Bamiyan and the Hindu Kush.

After burying the dead, the army moved out of Ghazni in two columns, the commander-in-chief riding at the head of the advance and Shuja following behind with Cotton. Eight uneventful days later, on 7 August 1839, the Army of the Indus stood massed at the gates of Kabul. Keane's dispatch to Auckland on this triumphant occasion is deliberately sparing of detail. Understandably so, for a true report of the reception accorded the conquering army would have caused Auckland some considerable embarrassment. 'The King entered his capital yesterday afternoon, accompanied by the British Envoy and Minister, and the gentlemen of the mission, and by myself, the general and staff officers of this army.'[32] Crisp, military prose devoid of the sort of bravado that had leapt from Macnaghten's report on the entry into Kandahar.

It was one of those rare historical moments that yielded an almost unanimous verdict by analysts. Shuja was mounted on a white charger, the king dazzling in coronet, jewelled girdle and bracelets. He was flanked by Macnaghten and Burnes. Bringing up the rear of the royal cavalcade came a squadron of the 4th Light Dragoons, another of the 16th Lancers and a detachment of Horse Artillery. Shuja had expressed his wish to enter Kabul in the company of British troops, with only a small number of his own Indian and Afghan soldiers in the procession. Goodness knows what was going through the emir's mind when he conveyed that tactless request to Keane. If Shuja sought to impress his subjects with a show of imperial British might, the crowds of onlookers remained singularly unimpressed. 'There was no voice of welcome', reported Kaye, a contemporary chronicler. 'The citizens did not care to trouble themselves so much as make him [Shuja] a salaam, and they stared at the European strangers harder than at his restored majesty.'[33] Kaye dismisses the victorious king's return as 'a dull procession'; with 'no popular enthusiasm, the voice of welcome was still'.[34] Kaye makes the telling point that those few who came to the thresholds of their houses to view the procession stared at the British troops more than at Shuja. Durand was a personal witness on that day to how Shuja 'was received in his capital without a show of welcome or enthusiasm',[35] while for another writer, the ceremony transpired 'entirely without a show of welcome'.[36]

What were they thinking, Shuja, Burnes and Macnaghten, each destined to die by an assassin's hand, as they rode side-by-side through the sullen city? Shah Shuja was securely cocooned in his phalanx of British soldiery. As with tyrants throughout history – Mussolini, Ceausescu and Saddam Husain in our time – Shuja could never conceive of himself one day being cut down by his adoring subjects. And what of Burnes, who was to be the first to fall? Was he perhaps wondering if the ghost of the change of heart towards Dost Mohammed had come to haunt him? Macnaghten, as one might imagine, remained blissfully indifferent to the realities passing before his eyes. The soldier-diplomat Lieutenant Colonel Sir Kerr Fraser-Tytler rightly asserts that Macnaghten 'was proof against all warnings'.[37] The envoy soon busied himself with correspondence to Auckland concerning the translation of Shuja's adulatory letter to Queen Victoria, the emir's decision to bestow a new military honour to be known as the Order of the Durrani Empire, and the critical question of whether this distinction was applicable to soldiers who had not taken part in the siege of Ghazni and the occupation of Kabul.

The logical place to garrison an occupying force was the Bala Hissar, the sprawling citadel that dominates Kabul. Its hilltop location, plus its 20-feet-high and 12-feet-thick walls, made it all but unassailable. Lieutenant Vincent Eyre, who was attached to the Bengal Artillery, said in his memoirs, 'Every engineer who had been consulted since the first occupation of Kabul by our troops, had pointed to the Bala Hissar as the only suitable place for a garrison which was to keep in subjection the city and the surrounding countryside.'[38] Shuja was intent on exerting his newly-acquired powers: the British might be his protectors, but make no mistake, ultimate authority resided in the monarch.

Macnaghten bowed to Shuja's wishes, unwilling to give an appearance of permanent occupation, which is how the people of Kabul would have interpreted a move en masse to the Bala Hissar. The army engineers had urged Macnaghten to override Shuja's wishes and to billet the troops in the citadel. Captain John Sturt, the army engineer in charge of public works, recommended the Bala Hissar be occupied at all costs. Even so, the envoy put aside all objections to avoid a conflict with the emir. Divided counsels and friction between the civil and military power were to be a fruitful cause of disaster from the outset. Shuja argued that the Bala Hissar overlooked not only his capital, but his royal palace as well. Furthermore, where else was he expected to house his wives and retainers, who comprised an entourage of more than 800?

The army had spent a rough winter in 1839 camped in the snow outside the city walls. The men were clamouring for a roof over their heads and a hearth by which to warm themselves. A spot of open ground was selected to the north-east of the city, roughly two miles from the Bala Hissar, and this became known in the ranks, not without a twinge of bitterness, as the 'Folly on the Plain'. If pushing an ill-equipped force hundreds of miles across the Sind Desert amounted to a near-fatal tactical error, the Kabul cantonment was a worthy follow-up in strategic mismanagement, and one that was to seal the garrison's eventual fate. Eyre took one look at the site and exclaimed that it was a wonder any set of officers in a half-conquered country should 'fix their forces . . . in so extraordinary and injudicious a military position'.[39]

The cantonment stood, in defiance of rule and precedent, on a low piece of swampy ground commanded on all sides by hills and forts that lay in Afghan hands. The mission compound was erected at one end of the 1,000-yard-long low rampart enclosing the cantonment. A complex of buildings served as the residence of Macnaghten and the senior officers. This exposed compound further weakened the cantonment's defences, for it had to be defended in time of siege. But the most unaccountable blunder of all, which gave the entire undertaking more the appearance of a hara-kiri mission than the construction of a fortification, was having the commissariat stores in an old fort detached from the cantonment. Everything was subservient to economies, the handful of dissenting officers was told. The Government of India was bent on keeping costs to a minimum after the vast expenditures the army had incurred on its march across Afghanistan. So in August 1840, the troops settled into their indefensible cantonment, on a site now occupied by the US Embassy in Kabul.

After failing to adequately provision an army on the march and having erected an unprotected fortress for their defence in the name of belt-tightening, what more could the Government do to invite catastrophe? Once again, Macnaghten rose to the occasion. So confident was the envoy that the army was fully in control of Kabul, he issued instructions allowing the officers to send to India for their wives and families.

First and foremost of the wives was Lady Frances Macnaghten who, with her collection of cats, sat on the top perch of the pecking order. But certainly the most formidable woman in the convoy trundling up from India was Lady Florentia Sale, the wife of Brigadier Sale, who was to earn great fame as the defender of Jalalabad. Truth be said, when the garrison was looking catastrophe in the face it was the women, and Lady Sale foremost amongst them, who showed a magnitude of valour conspicuously absent in the officers' ranks. General Sir Charles Napier himself was later to thunder, 'God forgive me, but with the exception of the women, you [the military commanders] were all a set of sons of bitches.'[40]

For the Afghans, the ladies' arrival at Kabul signified something quite distinct from a simple reunion of families, and moreover with husbands who had been freely cavorting with their own womenfolk at every opportunity. This was taken as a clear sign that the army was not contemplating an early departure, despite Macnaghten's assurances to the contrary and what was explicitly pledged in the Simla Manifesto. It was in all a rum set of circumstances, given Macnaghten's earlier insistence that Shuja be allowed to take possession of the Bala Hissar precisely to avoid giving the impression of a long-term British occupation.

For now, however, jolly times were in store for the garrison. The Afghans contemplated in fascination the spectacle of couples ice-skating on Kabul's frozen lake, and sailing across it in the summer. A cricket pitch was rolled out, there were tea parties, amateur theatricals and sumptuous banquets at the mission compound, all in the classic tradition of a British colonial outpost.

Then in October 1840, the fugitive Dost Mohammed reappeared from the Hindu Kush at the head of some 6,000 Uzbek Muslims who had flocked to his standard. To the army's great alarm, even some of Shuja's troops had deserted to join forces with the great deposed emir. The tribal column rode to Parwandarah, in the hills some 50 miles north of Kabul, where on 2 November 1840 they were intercepted by a force sent out under Sale. Dost Mohammed sat majestically on his horse atop a hill and exhorted his men to follow him 'or all is lost'. Then he gave the signal to move forward and led his cavalry to the attack against the British column. With banners unfurled, the Afghans cut their way through the terrified native troops to nearly within reach of the British guns. The Indian soldiers failed to stand their ground and were scattered by the enemy. The British held firm and paid a terrible price, with three officers killed and two severely wounded. Yet before the import of this debacle had sunk in, Dost Mohammed managed to snatch defeat from the jaws of victory. Inexplicably, the Barakzai leader rode into Kabul accompanied by a lone attendant, where he gave himself up to Macnaghten. Perhaps it was a reluctance to prosecute a war against his former friends (Burnes was with the troops), or that his honour had been satisfied at this last stand, or the acknowledgement that to carry on fighting against a superior force would in the end only bring greater humiliation to his people. We shall never know Dost Mohammed's motivation for surrendering, but his sword was warmly accepted by Macnaghten, who came to meet him by the banks of the Kabul River. The envoy arranged for his removal under escort to Ludhiana, which was gaining a reputation as the British abode for unemployed Afghan rulers, and there he remained in exile for two years.

It was now time for the Army of the Indus to be broken up. The entire Bombay column was ordered to retrace their steps via the Bolan Pass. A portion of the Bengal Army left for Peshawar by the route of Jalalabad and the Khyber Pass. Two key cities, Kabul and Kandahar, were to be held by British and Indian troops, along with the principal posts on the main roads to India, namely Ghazni and Quetta in the west, and Jalalabad and Ali Masjid in the east. It is worth highlighting the fact that communications were almost non-existent between these outposts, another fatal flaw in the army's strategic planning. Sale would keep a brigade in Afghanistan, Keane was to return by the Khyber, while Major General Sir Thomas Willshire marched westward to the Bolan Pass. Apart from Shuja's contingent, the garrison left behind in Kabul consisted of four infantry regiments, two batteries of artillery, three companies of sappers, a regiment of cavalry and some irregular horse, in all a force of some 1,500 British and 4,500 Indian sepoy fighting men, fully equipped and in good order. A Sikh contingent at last came up to Kabul, led by Lieutenant Colonel Sir Claude Wade, who brought with him Shuja's son Timour. It was a token force at best, reluctantly assembled after Ranjit Singh's death. Its contribution to the Afghan campaign was nugatory, to the degree that Kaye thought fit to describe it as 'absolutely contemptible'.[41] Within a few weeks Keane was back in India and had been created a baron, Auckland was made an earl, Macnaghten a baronet, and a shower of lesser distinctions descended upon the subordinate officers. Macnaghten and Shuja moved the court down country to the warmer clime of Jalalabad until mid-April to escape Kabul's severe winter, and apart from a scattering of tribal risings that needed to be taken in hand, with the Ghilzai Pashtuns up in arms in Kohistan and the Afridis causing mischief in the Khyber, a lull settled over Afghanistan and its snowbound capital.

In spite of the good times that had been enjoyed by all,

no regrets were heard in the ranks at the prospect of leaving Afghanistan. As he was preparing to depart Kabul, Keane turned to a fellow officer who was to accompany him on the march and uttered the fateful prophecy, 'I cannot but congratulate you on quitting the country, for mark my words, it will not be long before some signal catastrophe takes place.'[42]

CHAPTER 3

The Present Happy Moment

The hour had arrived for a new cast of characters to make their appearance on stage. The Old Guard began to disperse in late 1839, with the return of part of the Army of the Indus to India. Cotton had for some time been complaining of ill health, in all probability associated with his extreme corpulence and the rigours of performing military duties in an alien environment. In October, he relinquished the command of the Bengal Army to take up a less taxing post in the Bengal Presidency.[1] Macnaghten's relations with the military were always abrasive and strained, and never more so than in his dealings with Cotton. With a great sense of relief, the general was soon to extricate himself from this daily source of friction, for the good-natured but slow-witted Cotton was no match for Macnaghten's acerbic character. Brigadier Abraham Roberts, father of 'Bobs' Roberts, the hero of the Second Afghan War, gave up his command of Shuja's contingent. Roberts was another who found himself weary of Macnaghten's obstinacy, the envoy seeming to scatter stumbling blocks in his path at every turn. The final straw came in a row over Roberts's proposed plan to build an Afghan national army, an undertaking which, if successful, would have allowed the British to disengage from Afghanistan at an early date. Auckland could not afford to alienate his right-hand man in Kabul. When the governor general learnt of this antagonism between Macnaghten and the general, he treated Roberts's complaints as an offer of resignation. Roberts's replacement was the more pliable Brigadier Thomas Anquetil, an old India hand who was pushing sixty.

Macnaghten considered his position to be unassailable: he had faced down the 'croakers' in the army who had been opposed to the construction of the cantonments, he had dispatched at least two senior officers who were unconvinced by his cocksure optimism concerning security in Kabul, and now he was to reap the rewards of his unerring political leadership. Macnaghten's glowing reports from Kabul earned him the esteem of the mighty Hobhouse, President of the Board of Control, who informed Auckland that the envoy was to quit Kabul to take up the post of Governor of Bombay, one of British India's most prestigious civil appointments. Macnaghten replied with breathtaking arrogance that 'everything was quiet from Dan to Beersheba'.[2] He rejoiced in the knowledge that he was about to depart Afghanistan for good, with no burden of anxiety or fear, never imagining that he was not destined to leave the country alive.

It is now time to meet the two characters who, along with Macnaghten, were the principal agents behind the Kabul garrison's downfall. The designation of Major General William Elphinstone as successor to Cotton as commander-in-chief in Afghanistan must rank as one of the worst strategic blunders in British Army history. Elphinstone was described by the pre-eminent army historian John William Fortescue as being 'so infirm, so much crippled by gout and in such miserable health, that he was quite unfit for any work. He suffered intense and constant pain and was, to all intent, a dying man.'[3] There was never any dispute over Elphinstone's credentials as a courageous and capable soldier. He had commanded the 33rd of Foot, the Duke of Wellington's old regiment, at Waterloo with such distinction that he was made a CB, or Companion of the Order of Bath, the fourth most senior of British orders of chivalry.

A quarter of a century later, this crippled officer was painfully making his way towards Kabul in a palanquin, one arm in a sling, to take charge of a situation about to explode to pieces. Fraser-Tytler regarded Elphinstone as 'the worst possible man to exercise the effective leadership required at the doomed Kabul garrison'. Elphinstone, the British diplomat concluded, was 'too infirm in mind and body and ignorant of the country or people with whom he was dealing'.[4]

This soldier was clearly in no fit state to command in Afghanistan, so who had sent him there? Auckland was for once not the mastermind of this gaffe, though as always he treaded cautiously through turbulent political waters and at no time raised any objections to the campaign to appoint Elphinstone to Kabul. The governor general and his family were long-standing acquaintances of the old general. They were on such intimate terms that Auckland's sister Emily affectionately dubbed him 'Elphy Bey' (*bey* is the Turkish term for 'commander') in her diaries. It is sad but telling that even she should share the widespread pessimistic judgement regarding Elphinstone's failings.

When Auckland heard that Cotton was retiring, he brought the matter to the commander-in-chief, India, Sir Jasper Nicolls who, to the astonishment of many, saw in Elphinstone a man worthy of the task 'if his health should be such as to enable him to undertake such a command'.[5] It beggars belief that Nicolls could have been ignorant of Elphinstone's decrepitude, but he was most certainly aware of influence-wielders at home who were flexing their muscles over this debate. One person who was instrumental in securing Elphinstone's appointment was Fitzroy Somerset, a man of questionable wisdom who, as the future Lord Raglan, bore responsibility for the charge of the Light Brigade at Balaclava. Somerset was military secretary at the Horse Guards, a position that gave him considerable authority in the matter of army commissions. Hobhouse, on the other hand, had expressed serious reservations about this appointment. But Somerset was another of Elphinstone's friends and moreover, he had an open door to the Duke of Wellington's office and was able to persuade the Iron Duke[6] to turn a deaf ear to Hobhouse's scepticism. Wellington was seventy and soon to retire from political life. By his own admission, he was weary of Indian affairs and had little interest in being drawn into squabbles over military appointments.

So it came to pass that William George Keith Elphinstone was dispatched to Afghanistan. In the end, Auckland was delighted to have Elphinstone shipped out to Kabul. On the one hand, he would be easy prey for Macnaghten, the man who really mattered to the governor general. There was also the question of if not Elphinstone, then who? On this subject, most of the senior officers were of the same mind: Brigadier Sir William Nott was the best man on hand. But the very thought of Nott assuming command in Kabul would have sent a shiver down Auckland's spine, while Macnaghten would be certain to turn apoplectic. Here was a soldier to be feared as well as respected. Soon after entering Afghanistan with the Army of the Indus, Nott was ordered to the provinces, to command the troops at Kandahar, a comfortable 284 miles from Kabul. Nott did little to ingratiate himself with the political authorities. One of his first acts on arriving at Kandahar was to have a number of the irregulars under Prince Timur, Shuja's son, publicly flogged as punishment for robbing some of the city's residents. Cotton stood by his fellow officer, while Macnaghten was outraged and Auckland expressed his 'great regret and displeasure'. This one rash act effectively put paid to Nott's hopes of taking over command in Afghanistan. 'With this, and other brushes with authority in mind, Auckland did not consider Nott as a possible successor to Cotton.'[7] Nott was a fine soldier of the Old School, fearless, intolerant of half-measures, independent-minded – and unwanted. As commander-in-chief he was the one man who might have saved the garrison from destruction.

The spotlight now shifts to the doddering general's second-in-command. Elphinstone struggled to contend with Macnaghten the despot, who from the outset took pains to impress upon the kindly old man that the army was expected to dance to the tune played by himself and the 27 other political officers stationed about Afghanistan. The last person Elphinstone needed as his number two was someone of a morose nature and unpredictable temper, a person disposed to nurse grievances and who was disrespectful of his senior commanders and roundly disliked, even hated, by all who served under him. This was Brigadier John Shelton. Macrory sums up Shelton's two qualifications for the post: 'Shelton had . . . long service in India and a good measure of physical bulldog courage.'[8] One could stack against these qualities a litany of good reasons for keeping Shelton as far away as possible from Kabul.

Shelton's army career resounds with feats of valour. He was a veteran of the Peninsula War, taking part in the siege and capture of Badajoz as well as in four other major battles, including the storming of San Sebastián. It was in this latter campaign that Shelton lost his right arm, and it was said that he stood outside his tent, unmoved, while the surgeon amputated the shattered limb. He was a man of exceptional courage, conspicuously so in the desperate fighting at the Arakan offensive in the First Burmese War of 1824–6. He commanded the 44th of Foot in India and at the end of 1840 he was put in charge of a brigade to replace part of the force in Afghanistan. It is hardly surprising that Shelton's attitude towards his senior officer in Kabul was one of utter contempt. When the inevitable uprising broke out in Kabul, Shelton would grudgingly attend Elphinstone's councils of war, where he made himself as disagreeable as possible and rarely replied to the commander-in-chief's questions. Instead of having Shelton disciplined for insubordination, Elphinstone was moved to whimper, 'Shelton seemed to be actuated by an ill feeling towards me.'[9]

Elphinstone found himself caught between two powerful adversaries, Macnaghten and his insatiable ambition on the one hand, and the imperious, defiant Shelton on the other. It didn't take long for the general to fall victim to the former's colossal ego. Elphinstone struggled to comprehend the confused situation in Afghanistan, with word coming in almost daily of outbreaks of tribal unrest in the hinterland. The army had to cope with a revolt in Kelat, reverses sustained in Sind, attacks on Quetta, and other rebellious activity that threatened stability in Afghanistan. Macnaghten appeared at his residence one day in early June demanding that he order Shelton's approaching brigade to be diverted to Peshawar by forced marches. The reason? The multitude of Shuja's zenana, which was wending its way upcountry from Ludhiana, had been waylaid by a mob of mutinous Sikh soldiers. The ladies needed to be saved from these brigands. Elphinstone was not so weak-minded as to fail to perceive the folly of sending a badly needed brigade to rescue Shuja's harem. But the general was incapable of standing up to the envoy, so the orders were issued for Shelton to turn his column round and make for the Indus. Shelton obeyed, the Sikhs fled in panic at the approach of battle-seasoned British troops and on 10 June 1841, the brigadier led his men into Kabul at the head of a long train of women and their retainers. For Shuja, the safe arrival of his legion of wives came as something of a mixed blessing. Within a few weeks, he was negotiating loans with the bankers of Kabul to enable him to support his vast harem.

Meanwhile, 3,550 miles away, policies were being devised that would have a momentous impact on the unsuspecting Kabul garrison. In July 1841, the Conservative leader Sir Robert Peel was returned to a second term as prime minister. Peel's ministry came to power in the depths of a severe recession and a slump in world trade, with the Government shouldering a £2.5 million deficit run up by the previous Whig administration under Viscount Melbourne. Cutbacks and retrenchment were the words that echoed through the corridors of Westminster in the autumn of 1841. The Government noted with alarm that the Afghan expedition was costing the Indian revenue well over £1.5 million a year, more than half the national deficit. Macnaghten found himself under tremendous pressure to cut expenditures, which incidentally scotched Elphinstone's highly sensible suggestion to shift the cantonment to a more easily defensible site. 'The home authorities', Kaye tells us, 'had written out urgent letters regarding the miserable results of the continued occupation of a country that yielded nothing but strife.'[10] Hobhouse was warming to the view that the preferred course of action was to write off the Government's losses by withdrawing from Afghanistan. The President of the Board of Control feared that maintaining Shuja on his throne and keeping the peace in Afghanistan would require the deployment of a much larger force and a commitment to keep it in the country for many years to come. Auckland's response was to raise money by issuing debt at 5 per cent to cover Afghan expenses. Macnaghten was so blind to reality that when he learnt of the new loan being opened, he assumed it was to finance the Opium War in China. When the truth eventually sank in, that it was necessary to shore up the drain on the resources of the Government of India, he took the first of two disastrous decisions that set the tribes in open revolt.

Macnaghten was making preparations for his departure to India for the end of October, where he was to take up his post as Governor of Bombay. Lady Macnaghten had even begun hosting tea parties at their residence, at which friends were invited to view the couple's furniture up for sale. The envoy's anticipated departure brought a ray of hope for the confrontational garrison, for next in line in the hierarchy was Alexander Burnes, as we have seen a man with no shortage of ego but with far more knowledge of Afghan realities and lacking Macnaghten's inveterate bully instinct. Moreover, Burnes had many admirers in Government circles, in which his opinions on Central Asian affairs were highly valued. Another piece of good news was that Macnaghten would be taking with him to India poor old Elphinstone, who almost from the day of his arrival in Kabul had been begging to be relieved of his duties on grounds of failing health. At one point he pleaded to Macnaghten, 'If anything were to turn up I am unfit for it, done up body and mind, and I have told Lord Auckland so.'[11]

The best news of all for the army was that Nott had been chosen as Elphinstone's replacement. This marked an almost unprecedented departure from tradition, for it was rare indeed for an East India Company (Indian Army) rather than a Queen's Commissioned (British Army) officer to be given such a high-ranking command. The deteriorating situation in Afghanistan highlighted the need to bypass conventional elitism and bring in a soldier whose ability to deal with a crisis was above doubt. Nott, who frequently complained of being marginalized by army snobbery, was delighted at the prospect of taking charge in Afghanistan.

Macnaghten knew perfectly well that the step he was about to take would not go down well with the tribes. It was a great risk, calculated on the hope that the gratitude he earned from the Government would outweigh the repercussions that might follow, mainly from the warlike Ghilzais.[12]

The tribesmen collected £6,000 a year in bribes, officially designated as 'subsidies' or 'allotments', in exchange for undertakings to refrain from mischief in these strategic passes between India and Afghanistan. A contemporary report in the India Office archives records the success achieved with this proactive system:

> During the period of nearly three years in which they have been in our pay, not one single letter or dispatch was ever lost on its way by this route to Kabul, and the transit of the whole merchandise passing the hills was as secure as if in one of our own provinces.[13]

Macnaghten summoned the tribal chiefs to his residence one evening and treated them to an extensive lecture on the Government deficit. There was a pressing need to cut expenditures, he explained to his stern-faced audience, who squatted on the floor, sipping their tea in silence. Then he dropped the bombshell: the annual subsidies were to be halved to £3,000 a year. When Macnaghten had ended his exposé, the chiefs nodded their understanding and rose to leave as one, not uttering a word. Macnaghten looked on in dismay as the bearded Afghans filed out the door, with the knowledge that they had the power to close the passes with the ease of switching off a tap. After taking a solemn oath on the Koran to exact revenge on the traitorous Feringhees, this is precisely what they did.

Macnaghten's second cost-cutting blunder was to dispatch Sir Robert Sale's brigade back to India, thereby reducing by half the Kabul garrison's fighting strength. In early October 1841, the brigadier was ordered to be in readiness for the march. His Queen's 44th Brigade was a formidable force, consisting of the 13th Light Infantry of 800 bayonets, with Sale in command, and the 35th Native Infantry of roughly the same numerical strength, under Lieutenant Colonel Thomas Monteith. The homeward-bound march was expected to be peaceable, for this is what the men were told by their commanding officers, who were acting on intelligence received from Macnaghten and his political officers in the field. It was just as well that no trouble was anticipated on the road eastward to Jalalabad, for the men were equipped with inferior weapons, mostly Baker and Brunswick muskets with an accuracy of up to 300 yards, less than a third the effective range of the tribesmen's long-barrelled *jezails*. Sale was aware that these old flintlocks were unreliable and just as likely to misfire as to shoot wide of the mark. He also knew that the army had in its Kabul arsenal 4,000 new detonator-type muskets. The brigadier begged permission to arm his regiment with these more modern weapons, but Elphinstone turned down the request. The general reasoned that the soldiers had no need of new muskets, as they were departing on the first stage of their journey back to England. The later consequence of this decision was the loss of about a quarter of the weapons in store to the Afghan rebels. The 35th moved out on 9 October and we shall later see what fate befell them on the elevated plain that separates Kabul from the Bootkak hills.

Back in Kabul, the days of cricket and dances were rapidly drawing to a close. In a word, the party was over. The second day of November 1841 fell in the middle of the month of Ramadan, a time in which Muslim tempers are traditionally on a short fuse. It was also the day following Macnaghten's meeting with the tribal chieftains. Kabul awoke to the sound of gunfire and tumult that morning. Macnaghten's wake-up call, in every sense of the word, came in the form of an urgent note from Burnes reporting the gathering of an angry mob outside his house. The insurgents had begun attacking the residence as well as the treasury, though in the first moments of confusion, Burnes never suspected that a full-scale insurrection was in progress and he ventured to his balcony to harangue the crowd. A volley of fire from the streets quickly removed any doubt that he was dealing with something more than a petty riot. Shuja could see the attack from his palace and duly dispatched a regiment to disperse the crowds, but his men got pinned down in the narrow streets and were driven back, lucky to escape a wholesale massacre. Shuja had little idea of what transpired outside the Bala Hissar compound. Proof of this was a message he sent across the city to Macnaghten, reassuring the envoy that as far as he could tell, all was still well with Burnes, when in fact his disembowelled corpse was already hanging from the branches of a tree in his garden.

Throughout the brief siege, Burnes could not bring himself to believe that his life was under threat. It was an altercation, he preferred to think, perhaps a protest over the misbehaviour of some drunken soldiers, or a pay dispute. If things got out of hand, surely Macnaghten would be dispatching a column in short order to disperse the noisy swarm of tribesmen. At least, this is what he had requested in a hastily-penned note to the envoy. Burnes had also dispatched two runners with messages to one of the tribal chiefs, Amanullah Khan, entreating him to call off his followers. The gravity of Burnes's plight sank in when only one of the messengers returned barely alive. The other had been decapitated at a single stroke by one of Amanullah's men. That was when Burnes rushed out to a gallery where, with his usual cockiness, he promised the tribesmen a handsome reward if they agreed to disperse and return to their homes. One of the officers, Lieutenant William Broadfoot, had already fallen to an assassin's bullet through the heart. Burnes's faithful Kashmiri servant Mohan Lal was saved from death after taking refuge with the household servants' wives.

At roughly the same time the insurgents were moving in on Burnes's house, Shelton was readying his men to storm the city, but the supremely sanguine Macnaghten ordered the brigadier to stand down, with the somewhat inaccurate news that quiet had been restored to Kabul. Burnes now realized he was staring death in the face, when suddenly a Kashmiri merchant burst into the room begging Burnes and his brother Charles to disguise themselves as natives and slip out with him to the garden to make their escape. The two Englishmen saw no alternative but to try this last gambit, and as they took one terrified step outside, the supposed Kashmiri merchant shouted 'Here is Sikander [Alexander] Burnes!' There was a flash of knife blades and a moment later, the two brothers lay lifeless on the ground. Rotted bits of Burnes's corpse still hung from the tree when the army began its evacuation of Kabul two months later. One wonders, when the Afghans sacked and destroyed what was left standing of the residence, what they made of Burnes's tins of hermetically sealed salmon, imported from Aberdeen, and the cases of finest French champagne laid down in his cellar.

Forbes argues that the 2 November uprising may not have been the result of a fully-organized plan. He writes, 'There are indications that it was premature, and that the revolt in force would have been postponed until after the expected departure of the Envoy and the General with all the troops except Shelton's brigade, but for an irrepressible burst of personal rancour against Burnes.'[14] There is no doubt that Burnes was despised by the Afghan leaders. Akbar Khan hated him for what he saw as Burnes's betrayal of his father Dost Mohammed. Many of the chieftains reviled him for allegedly plying his charms on the lovely ladies of Kabul. Burnes was known to be a womanizer, but there is no evidence of his having engaged in any dalliances with Afghan women. In any case, from a purely tactical standpoint, the tribesmen had gone for a soft target, as Burnes's house was nearer and less heavily defended than the cantonment.

There may have been another, more sinister motivation for Burnes being chosen as the first British victim of the uprising. It was well known that Burnes was detested by Shuja, since he was the Government agent who had argued against the Saddozai pretender's return to power. The tribal chiefs had paid a visit on Shuja after their meeting with Macnaghten, where they received the news of the cut in subsidies. They laid their grievances before the emir, who threw up his hands in despair, claiming that he was powerless to do justice to their cause. Shuja, it must be emphasized, was well known for cajoling the British with money matters, but he was less inclined to step in and make good on Macnaghten's financial commitments. His perverse train of thought would dictate that he had nothing to lose if his subjects turned against the British masters, who kept him imprisoned, like Prince Segismund, in a tower of dreams. He indeed had everything to lose, but he was late in picking up on this point. The contemporary French historian Joseph Pierre Ferrier travelled widely in Afghanistan in the years shortly after the First Afghan War and served as Adjutant General of the Persian Army. He had access to first-hand information from veterans of that campaign, though some of what he relates is undoubtedly coloured by the strained relationship which in that day existed between Britain and his country. Ferrier says that one of the Afghan chiefs in attendance in the meeting at the Bala Hissar was Abdullah Khan, who later met in secret with Shuja in his palace garden. 'It was resolved that the death of Sir Alexander Burnes should take place immediately', writes Ferrier, 'for the king, who well knew the infinite trouble that officer had taken to support Dost Mohammed, detested him, and was very anxious to prevent his being made the Resident at his Court, if Sir William Macnaghten, whose departure had been talked of for some time, should be removed.'[15]

Whatever the driving force behind Burnes's murder, the deed was done. But it can be safely assumed that the Afghans hadn't intended this to be a full-scale challenge to the British garrison. This is borne out by the fact that for 24 hours after the attack on Burnes's home, none of the ringleaders dared venture into the streets for fear of British reprisals. Winter was approaching and with the cold season came the snows that blocked the passes to India. It would have been unrealistic to expect the British to abandon Kabul at this time of year without putting up a stiff fight – though little did anyone suspect that this is precisely what they were to do a few weeks later. If the chiefs could keep the rabble in check, their best hope was to exact a pledge from Elphinstone and Macnaghten to evacuate the capital in the spring of 1842. The brutal murder of Burnes and his staff showed that the mob was not under anyone's control, not in Kabul on that morning, not on the fateful march to Jalalabad.

That evening, Macnaghten conveyed to Elphinstone an account of the day's tragic events. Hardly able to believe the uprising gathering pace before his eyes, he turned to the general for assistance, but Elphinstone was wholly incapable of meeting the crisis. It betrays the envoy's state of denial that he had failed to take any military action and waited several hours to acquaint the commander-in-chief with this dreadful intelligence. What is even more astonishing is the general's response *by letter*, between quarters only a few yards distant from each other. This cannot be ascribed to some quaint Victorian formality – it was madness of the very first magnitude. 'Since you left me', Elphinstone wrote, 'I have been considering what can be done tomorrow.' He then goes on to speculate on the possibility of reprisals. Elphinstone stumbled across the thought that it might be possible to break out of the cantonment and meet Shelton's force, 'but not without very great loss, as our people will be exposed to the fire from the houses the whole way'. Of course, the general could not be expected to foresee the terrible losses the army was to sustain on its forced evacuation of the city. Elphinstone must have been considering the bedtime hour at the end of a troublesome day, for he ends his letter with the melancholy words, 'We must see what the morning brings, and then think what can be done.'[16]

The uprising was not confined to Kabul. The day after the Burnes's murder, an attack broke out at the Charikar garrison 40 miles to the north. A regiment of Shuja's Indian sepoys, comprised largely of Gurkhas, had been deployed to keep the tribesmen in check in the rugged Kohistan hill country. They were garrisoned in a small fort near the town, along with a contingent of women and children. The fort was three miles from the headquarters of Major Eldred Pottinger of Herat fame, now the local political agent. The bulk of the regiment was made up of young recruits who had never heard a shot fired in anger. When the tribes rose against them, they found themselves holed up in vulnerable barracks that were still under construction. The Gurkha adjutant, Colonel John Haughton was, along with Pottinger, the only person from the Charikar garrison to reach the relative safety of Kabul. On his arrival at the cantonment, he described to Macnaghten the perilous state the troops had faced, thanks to a lack of proper planning. He told the envoy it was apparent to all that in the event of an attack, the lack of water would be a great difficulty, but intelligence reports from friendly tribesmen had led him to believe such an event was impossible. Far from it. The attack on the Charikar fort began at daybreak on 3 November. Less than a fortnight later, the only two British survivors of Charikar rode into Kabul in the dead of night, narrowly avoiding the suspicious eyes of the insurgents. Pottinger was badly wounded in the leg and Haughton had lost his right hand to an Afghan sword. What had transpired was that after a week of fighting, the garrison's water supplies were indeed depleted and, ten days later, after the gunners and most of the Muslim recruits had deserted, Pottinger took the decision to abandon the fort and make a dash for Kabul. The column rode out under cover of darkness, but when they reached the first stream the thirst-mad pack broke into a wild scramble for water and the force rapidly disintegrated. While Pottinger and Haughton struggled ahead along a remote mountain path, what was left of the regiment followed the main road. The column got within 20 miles of Kabul, where they were spotted by the rebels and destroyed nearly to a man. The last British officer alive, the regimental doctor, held on until three miles out of Kabul, where he was too was cut down.

Macnaghten and Elphinstone were for once in agreement that help needed to be sought to restore order in Kabul. 'Incapacity and imbecility among the senior officers was naturally followed by demoralisation and cowardice among the troops', according to one observer.[17] In reality, Elphinstone was by now completely at the envoy's mercy and it was Macnaghten, in counsel with several senior officers, who insisted that the commander-in-chief call in a relief force from the provinces. On the same day that Burnes fell, Elphinstone had taken a tumble from his horse whilst inspecting the guards, and he became bedridden. From that moment, the army in Kabul was without a leader. Elphinstone 'was compelled to rely on the information of others, and to seek the advice of those with whom he was associated'.[18] It was inevitable for him to fall into the hands of Macnaghten, the most powerful voice amongst his advisers, though in fact Elphinstone was inclined to be guided by the last speaker's counsel, even that of junior officers or anyone else who had advice to offer.

So it was agreed that the general would send out an SOS – but who could be expected to gallop to the garrison's rescue? After the Burnes affair, Shelton was locked in the Bala Hissar with Shuja's 6th Infantry, whilst the rest of the force was coming under daily attack from the Afghan rebels who held fortified positions on the hills commanding the cantonment. On 9 November, Shelton received orders to return as quickly as possible to the cantonment. He boldly broke out of the citadel with his men to make his way across the city, meeting almost no opposition from the insurgents, who were concentrating their fire on the cantonment. When the brigadier arrived at the gate he was welcomed as a conquering hero, but he scarcely took notice of the jubilation, for he immediately grasped the garrison's desperate position: the defensive perimeter could be run over 'with the facility of a cat' and so many soldiers were required to defend the ramparts that not a man could be spared to confront the besiegers outside the walls. Given the grim situation Shelton had to confront, it was small wonder that Elphinstone, far from receiving his second-in-command with joy, complained of the brigadier's 'contumacious' manner. The general longed for the return of Sale, whom he regarded as better disposed towards him. By now, Elphinstone's state of despondency was so deep that he confided to his diary, 'I was unlucky also in not understanding the state of things, and being wholly dependent on the Envoy and others for information.'[19] Whatever his state of mind or body, Elphinstone held the senior military position in Kabul. Shelton would have perhaps taken a resolute stand against the attackers, even given the limited resources at his disposal, but he was powerless to act so long as Elphinstone remained in charge and hence, the army was left in a state of paralysis.

The only person in the cantonment who seemed to recognize the gravity of the crisis was Brigadier Sale's wife Florentia. Lady Sale was 53 and the mother of 12 children (four of whom died in infancy) when she travelled to Kabul in 1840. The diary she kept during the siege and on the retreat is one of the most powerful as well as colourful documents to emerge from the Afghan catastrophe. It provides a blow-by-blow narrative, dripping with sarcasm and ridicule, of a set of officers she collectively dismissed as 'croakers'. With good reason this formidable woman was dubbed the 'Grenadier in Petticoats'. As one scholar charmingly described her,

> When she stands on the battlements and watches the engagements outside . . . she is not sighing and fearful, as the poets portray the wives of Troy, watching the events unfold from the towers above. Rather, she is dodging the bullets, observing closely the order of battle.[20]

No sooner had the news of Burnes's assassination rocked the cantonment, than she wrote, 'It appears a very strange circumstance that troops were not immediately sent into the city to quell the affair in the commencement, but we seem to sit quietly with our hands folded up and look on.' For the men in charge of the garrison's defence, she reserves nothing but contempt: 'The state of supineness and fancied security of those in power in cantonments is the result of deference to the opinions of Lord Auckland, whose sovereign will and pleasure it is that tranquillity do reign in Afghanistan.' The most chilling entry of all was her foreboding of what lay ahead for the garrison: 'Most dutifully do we shut our eyes to our probable fate.'[21]

As if to confirm her scathing censure of those in charge, rather than sending out 4,500 well-equipped soldiers to rout the insurgents, Macnaghten dispatched a series of urgent appeals for help, first to Sale, who was presumed to be somewhere down country en route to India, and then to Nott, asking him to send a column from Kandahar. The beleaguered Kabul defenders, however, were out of luck on both accounts: Sale had more than enough to contend with on the road to Jalalabad to spare a thought for the plight of his comrades in the capital. Nott, for his part, was engaged in sharp fighting with the tribes around Kandahar, and with winter closing in, the road to Kabul would soon become all but impassable.

Sale instructed Colonel William Dennie, one of the heroes of the storming of Ghazni, to make for Jalalabad and secure the place in readiness for the attack that no one doubted would be coming from the eastern hills they had just escaped. George Gleig, the brigade chaplain, described the preparations:

> Strong working parties were employed from morning till night in filling up the breaches in the town walls ... every tree which stood in the line of fire was cut down ... every wall and house and inequality in the ground levelled.[22]

Parapets were run up along the ramparts, foraging parties were sent out to gather grain, sheep and fuel from nearby villages – in short, a textbook model of how to fortify a garrison under siege conditions, quite contrary to the ramshackle pile of a cantonment that had been thrown up in Kabul.

Sale, Dennie and the rest of the senior officers now glumly contemplated the latest letter from Macnaghten that had come in along with the report of the Kabul uprising. The envoy had finally resigned himself to the reality that the rebellion posed a threat to the garrison's very survival. With only a few days' food and fuel in store, Macnaghten took pen in hand to entreat Sale to return in all haste to Kabul to relieve the garrison. 'Our situation is a very precarious one', he wrote. 'We may be said to be in a state of siege . . . we have provisions for only ten days.' There then followed a frantic appeal to Sale's sense of duty, couched in the envoy's customary arrogance, in which he implored the brigadier to come to the rescue of Kabul 'if you have any regard for our lives or for the honour of our country'.[23] Macnaghten would have done well to reflect on his own lack of regard for his countrymen's well-being, by having vetoed proposals to take over the Bala Hissar, or refusing to allow a proper fortress to be built for the garrison, or failing to consider the repercussions of slashing the tribal subsidies.

After much soul-searching and consultation with his senior officers into the late hours in a series of councils of war, which often threatened to break into open confrontation, Sale resolved to stay put in Jalalabad and abandon his colleagues in Kabul to their fate. It was a controversial decision, to say the least, and one that still incites heated debate in military circles. Was Sale justified in disobeying orders, or is that irrevocably a contradiction in terms? It can be argued that by the time Macnaghten's letters had reached Jalalabad, the Kabul garrison's fate had been sealed. But Sale and the others in his command were ignorant of this development, just as Macnaghten and Elphinstone were unaware of the threat to Jalalabad, a town under siege. Sale saw no possibility of getting his brigade back through the passes alive. Akbar Khan had called for all-out jihad against the infidels, vowing to slaughter any Englishman who ventured into the hills, and he had the resources to carry out his threat. Sale would have been obliged to leave 300 sick and wounded behind in the care of native guards, whose loyalty was at best dubious. The brigade lacked sufficient transport and was short of ammunition. Numerous factors argued against Sale sending his men on what was assuredly destined to be a suicide mission. Yet there is no doubt that from a strictly professional standpoint and leaving aside the moral implications, Sale stood in flagrant dereliction of duty. This refusal to obey orders could have brought him up for a court martial, or worse. It was a terrible dilemma for the twice-wounded, ageing soldier to endure. It was a courageous decision and History has fortunately treated it as the right one to have taken.

The result of Sale's reply to Macnaghten was to drop squarely into the envoy's lap the terrible choice of whether to hold out in Kabul for the winter, or negotiate with the rebels the terms of the garrison's retreat to India. Of course there remained the third alternative of making a dash for the Bala Hissar. This option dominated the councils of war held throughout November 1841, as the British contemplated the probability of an early fall of snow. An adverse turn in the weather could greatly, and perhaps fatally, slow the progress of a convoy of troops and their baggage train along the two-mile trek from the cantonment to the security of the citadel. Macnaghten at first expressed himself strongly in favour of abandoning the cantonment – Shuja's feelings regarding this matter had suddenly been assigned a lower priority. The military authorities, in their dithering way, put forth many objections to this plan in order to overrule the envoy. Lieutenant Eyre was one of the army's few dissenting voices: 'I venture to state my own firm belief that had we at this time moved into the Bala Hissar, Kabul would have still been in our possession.'[24] The chief arguments against this proposal were the difficulties of conveying the sick and wounded, the want of firewood and forage, the humiliation it would have represented in the enemy's eyes, and the risk of defeat on the road to Kabul. Each of these arguments could easily be refuted by simple common sense. It might prove difficult, but by no means impossible to carry the sick, and the army had procedures in place for just such an eventuality. Firewood was in short supply but there was enough for cooking, which in itself provided warmth and in far more comfortable surroundings than the cantonment. The horses would need to be shot and added to the food stores – but what need had the garrison of cavalry behind fortified walls? A scorched-earth policy could have been applied to the cantonment, leaving nothing for the Afghans to gloat over. Finally, half the two-mile distance between the cantonment and the Bala Hissar was protected by the guns of the citadel, and if the cantonment's surrounding hills had been cleared of insurgents, the army could have likewise placed guns there to sweep the plain on the cantonment side.

A final ray of light lingered on the horizon before Macnaghten and the military would have to take this fateful decision: Nott was in full strength in Kandahar, which at a smart pace lay less than a fortnight's march from Kabul. Eyre recounted that Sale's expected return had brought grounds for hope:

> Our disappointment was therefore great to learn that all expectation of aid from that quarter was at an end. Our eyes were now turned towards the Kandahar force as our last resource, though an advance from that quarter seemed scarcely practicable so late in the year.[25]

Just how impracticable an undertaking this was is something they were shortly to find out.

There was yet hope for salvation, hence Macnaghten had changed his mind entirely about a dash to the Bala Hissar. He was now of the view that if provisions could be secured, all units should remain quartered in the cantonment. The dilemma was that the commissariat had been lost to the enemy, almost inevitably, given its poorly-guarded position outside the perimeter wall. When the attack on the commissariat came on 5 November, the officer in charge of the stores called repeatedly for reinforcements. 'Elphinstone, in lieu of reinforcing him, endeavoured to withdraw the garrison [from the commissariat], sending out several detachments to effect this suicidal measure', recounted Durand.[26] In a hastily-assembled council, the general was persuaded to see the folly of this rescue plan and the relief force was not ordered out of the cantonment. So with no help in sight the commissariat was given up without a struggle, to the immense delight of the hordes of Afghan looters who swarmed over the abandoned stores.

There was no let-up in the siege, with the enemy ranks having by this time swollen to the tens of thousands and giving no quarter. After ten days of relentless assault, on 22 November a detachment of cavalry was sent out to prevent the enemy taking the hilltop village of Bemaru, but they were beaten back with heavy casualties. The Bemaru action marked 'one of the most eventful and the most disastrous in the history of the insurrection'.[27] Shelton was incensed: Bemaru not only held a strategic position above the cantonment, it was also a storage site of badly-needed grain. Consequently, the next day at dawn an artillery barrage opened up on the village in an attempt to dislodge the enemy. This was followed by an infantry assault under the command of Major Stephen Swayne, who managed to lead his contingent in the wrong direction and took a bullet in the neck for his trouble. When Shelton spotted a body of several thousands of men approaching from Kabul, he swiftly ordered the troops back to the cantonment, but not before they had drawn up in two squares, Waterloo-fashion, thereby providing a perfect red-coated target for some of the world's finest marksmen.[28] The toll in men lost and wounded was terrible. Moreover, the troops had only one gun with which to defend themselves, in blatant disregard of army regulations that forbade fewer than two guns to accompany a force in the field.

By midday, after taking a pounding for more than five hours from the rebels, the Indian sepoys broke into a rout, as a stream of men came pouring down the hills towards the cantonment. There was not even an attempt to bring in the wounded, who were brutally cut up and mutilated where they lay. Eyre lamented: 'Our troops had now lost all confidence' and even the officers 'began at last reluctantly to entertain gloomy forebodings as to our future fate'.[29]

The siege continued to drain the garrison's dwindling resources and morale and then, on 10 December, Macnaghten received a letter that finally dashed all hope of relief from Kandahar. Nott had his hands full at Kandahar in repelling a threatened attack by the rebel chieftain Atta Mohammed Khan. The Afghan leader, one of the Kandahar sirdars, was outraged at seeing his homeland fall under the occupation of an army of infidels. Accordingly, he crossed the plains from Kabul at the head of a tribal force some 10,000 strong to attack Nott's garrison. The brigadier knew what was coming and he waited until the horde was within striking distance. Khan's force was taken by surprise and crushed in a furious 20-minute engagement.

While Nott and his men were recovering from their spectacular victory, on 14 November a letter from Macnaghten was brought into camp with orders to send a brigade in all haste to reinforce Kabul. To obey would mean a perilous reduction in strength of the Kandahar garrison, with the fanatical tribesmen in all probability massing for another assault on the hated Feringhees. Nott's position was different to that of Shelton's in Jalalabad. Nott enjoyed the good fortune of having as his political officer Major Henry Rawlinson, an Oriental scholar whose fame rested upon the deciphering of the Behistun cuneiform inscriptions in Persia. He was also a first-rate soldier who fought under Nott with distinction. Nott and Rawlinson saw eye to eye on matters of strategy, with none of the confrontation that plagued the relationship between Sale and several of his senior officers, mainly in military councils convened to debate the issues of whether to return to the relief of Kabul, or whether the garrison should attempt a breakout to Peshawar. Nott's force was up to divisional strength, while Sale commanded at most a couple of thousand men. Nott's decision to obey orders was influenced by his relatively secure position behind Kandahar's solid walls and with no enemies within, the brigadier having summarily expelled the city's residents. But it was a decision taken against his better judgement – he confided to his officers that the action was not of his doing and that he seriously feared the men were all going to their destruction. But he also reminded them that as a soldier, he was obliged to defer to superior authority.

Colonel James Maclaren was the officer chosen to lead the relief mission to Kabul. Maclaren had set off for India with the retiring native regiments on 8 November. The following day the column received the news of Burnes's murder and the Kabul revolt. Nott recalled Maclaren's brigade to Kandahar, whence it was re-equipped and sent off to Kabul. Within a few days, Maclaren's column was spotted on the plains, marching back towards Kandahar. The column had struggled to within a few miles of Ghazni, which sits at an icy elevation of 7,000 feet, where their progress was abruptly brought to a halt by heavy snowfall and freezing winds.

The curtain now rises on the final act of this drama, which unfolds like a formal Greek tragedy. 'We see the British in their prosperity, blind to their impending fate, the sudden reversal of fortune, the many opportunities to escape calamity all squandered through heedlessness and misjudgement, and in the end a final catastrophe.'[30]

The day after Macnaghten read the dispatch informing him that Nott's brigade had failed to make it through the treacherous snowdrifts blocking the road to Kabul, the envoy, with heavy but hopeful heart, arranged to meet the chieftains, amongst them the notorious ringleader Akbar Khan who had come down from Bamiyan. The task before Macnaghten was to negotiate an honourable withdrawal to India of the Kabul force, along with all the other contingents garrisoned in Afghanistan. Macnaghten had drafted a protocol on what he considered to be highly-favourable terms. The British would depart Kabul around 15 December 1841, Dost Mohammed would return to Kabul to reclaim his throne, Shuja would be allowed to accompany the British to India, if he so desired, and the British would be permitted to retain a Resident at Kabul. The Afghans cheerfully agreed to these terms, and then began adding fresh demands for more hostages, including Shelton, to be given up to ensure the army's evacuation of the country, along with an undertaking to hand over the British guns and ammunition to the insurgents, and so on.

The result of this whittling-down tactic was the delay, day after day, of the garrison's departure, a process that set Macnaghten's nerves on edge. A successful withdrawal on satisfactory terms would place a feather in the envoy's cap, seeing him seamlessly from the frozen wastes of Kabul to the warm comfort of the

Bombay governorship. With this heartening thought in mind, Macnaghten received a kinsman of Akbar Khan at the residence on the evening of 22 December, to hear the Afghans' definitive list of demands. The news was not bad: Shuja could remain as emir with Akbar Khan as his minister in a power-sharing role. Best of all, the British would be allowed to occupy the Bala Hissar until a practicable date of departure could be set for the spring of 1842. It was a most acceptable set of proposals, even if Macnaghten had to delay his arrival at Bombay. Such was the envoy's state of relief that he failed to ask himself the vital question: why should the Afghan insurgents, who literally held the garrison's back to the wall, be offering such favourable terms?

Macnaghten consented to a meeting with Akbar Khan the following morning, 23 December, to which he went accompanied by two of his assistants, Captains Colin Mackenzie and George Lawrence. Mackenzie was later appointed Brigadier General commanding in Hyderabad, while Lawrence went on to take chief military command of all British forces in Rajputana in the 1857 Mutiny. Macnaghten never made it back to the cantonment that day. He had requested Elphinstone to deploy two regiments of infantry and two guns, just in case a plot were afoot, as Mackenzie had that day warned. Even at this late date, there prevailed in the military command the same absence of rational judgement that had characterized the force since setting up camp at Kabul. Elphinstone refused to provide the escort, for fear of exposing the men to enemy attack. Macnaghten was exasperated by the commander-in-chief's bizarre reasoning, for after all, this was what soldiering was all about. Yet he went along confident that if the meeting were successful, it would be worth all the risk. He failed to take on board the risks that awaited if the meeting went wrong.

It all transpired in a few moments: the two parties met by the riverbank and after exchanging formal pleasantries, Akbar Khan withdrew the pistols Macnaghten had sent him as a gift the previous day and shot the envoy dead. The last words Macnaghten was heard to cry out were in Persian: '*Az burai kodar!*' ('For the love of God!'). Another of the party, Captain Robert Trevor, was dragged from his horse and hacked to death by Ghazi fanatics, while Mackenzie and Lawrence were bundled away as hostages, narrowly escaping the mob's fury. The Afghans now turned their rage on the envoy's corpse, quartering it like the carcass of an animal under the butcher's knife. To Lady Sale fell the sad duty of informing Macnaghten's and Trevor's wives of their husbands' assassination. 'Over such scenes I draw a veil', she wrote. 'It was a most painful meeting to us all.' She presumably omitted the gruesome details in her meetings with the widows, which were confided to her diary pages. 'All reports agree that both the Envoy's and Trevor's bodies are hanging in the public *chowk* [square], the Envoy's decapitated and a mere trunk, the limbs having been carried in triumph about the city.'[31] One detail that escaped Lady Sale's notice was that Macnaghten's severed head had been impaled on a gate. That evening, in a macabre gesture one thoughtful soul came out to replace his glasses.

The eighteen chiefs of the Afghan confederacy, with whom Macnaghten had negotiated the garrison's capitulation, now went on the offensive. They made it very clear that it was they, and not the hapless British officers, who were to dictate the terms of surrender. Major Eldred Pottinger was the second most senior civilian in Kabul, and so to the hero of Herat fell the grim responsibility of negotiating the retreat to India. Pottinger was still bed-ridden, recovering from the wound he had taken in his leg on his escape from Charikar. He made no pretence of being honoured by his promotion to chief political officer: 'I was hauled out of my sick room and obliged to negotiate for the safety of a parcel of fools who were doing all they could to ensure their destruction, but they would not hear my advice.'[32] Pottinger argued that to give up meekly without a fight was madness, as well as a disgrace to British arms. Rather than squander precious time and supplies discussing a disgraceful capitulation, Elphinstone and his staff ought to dispatch the entire force into the city to crush the enemy and, for good measure, blow the place to pieces. It was commonly held that Akbar and the chiefs had not the slightest intention of keeping faith. Their treachery would surely show its face the moment the army abandoned its defensive position behind the cantonment walls – if not before.

Christmas 1841 was not surprisingly a dismal occasion for the British in Kabul, a day which, faced with starvation, freezing temperatures and hordes of sword-waving Afghans at the ramparts, the phlegmatic Lady Sale shrugged off as 'far from cheering'.[33] The garrison awoke to indoor temperatures as low as 11°F, and finally some news regarding their departure for India. The Afghan conspirators had presented Pottinger and Elphinstone with the fourth and definitive draft of the terms of evacuation. Dripping with hypocrisy, the document began, 'That at the present happy moment, to put away strife and contention, avert discord and enmity, the representatives of the great English nation . . . have concluded a comprehensive treaty containing certain articles . . .'.[34] The articles to which this treaty referred levied further limitations on the weapons the troops would be allowed to carry with them on the march, as well as a demand that all of the garrison's treasure be handed over to the Afghans, and also that the married men, their wives and children, be given as hostages. Pottinger was furthermore obliged to send letters to Jalalabad, Kandahar and the other British garrisons, with instructions to immediately abandon their positions and return to India. Then, on Boxing Day, the chiefs returned to the cantonment with fresh demands for money, and that was when Pottinger put his foot down, insisting that there was no point in treating with such treacherous people. Pottinger was above all else a professional soldier who had been thrust into an uncomfortable political role. For him, the only honourable choice was for the garrison to stay put or fight their way to Jalalabad. The other senior officers thought differently: remaining behind in Kabul or giving battle to the Ghazis on an 80-mile trek through snowdrifts presented two alternative forms of suicide. Pottinger was thus overruled and the treaty was duly signed.

Epiphany 1842 dawned as truly a moment of sudden and great revelation to the 16,000 people massing for the march from Kabul. The soldiers, their families and the many thousands of camp followers accompanying them greeted the morning of 6 January with mixed feelings of yearning and terror. Whatever may lie ahead on the road to India, this tumult of shivering, half-starved people and baggage animals assembling on the bitterly cold fields outside the cantonment took heart in the thought that more than two years' of misery and suffering would soon be behind them. But no one could ignore the frightening thought of what might be lurking in those dark, towering defiles and on the wind-swept snowfields they were about to traverse.

The first contingent of the advance column, with Elphinstone at its head, marched out of the gates at half past nine that morning. The great throng of men and baggage lumbered forward in such a state of disarray that it wasn't until late evening that the last of the column found itself clear of the city walls. The retiring army was comprised of five infantry regiments, a regiment of Bengal Light Cavalry, several squadrons of irregular horse, six guns of the Bengal Horse Artillery, three guns of the mountain train and a group of sappers and miners. Elphinstone had in all under his command 4,500 fighting men, of which around 700 were British. Added to these were some 12,000 camp followers, besides women and children who were, in Fortescue's words, 'in themselves a serious, if not fatal encumbrance'.[35] The 44th took the lead under Brigadier Anquetil, followed by the main body led by Shelton, with the rearguard commanded by Colonel Robert Chambers. It was absolutely imperative to hustle the column on without delay. The first and most crucial task was to make it past the Khoord Kabul Pass on that day's march, a distance of 19 miles east of Kabul. With luck they might be able to reach Tenzin, which lay another ten miles along at a lower altitude, at which point they would be free of snow. Had it been a matter of only the men under arms and their families, the objective might have been achieved that very day. Fortescue adds that 'Shuja had finally decided to stay at Kabul, so that one useless encumbrance at least was out of the way.'[36] But even without the emir and his great retinue of wives and servants, the army's many thousands of camp followers in tow meant that putting the pass behind them in one day was out of the question.

Shortly after the march had begun, the Afghan chief assigned to provide the promised escort galloped up to Elphinstone, begging him to delay their departure until the following morning because his men were not ready. Whether this was a show of genuine concern or merely a ploy to cover the rebels' backs, we'll never know, but it was clearly too late to put a halt to the crowds of soldiers and followers streaming out the gates. As the last contingents plodded out of the cantonment enclosure, the plunderers moved in on the rearguard, picking off the terrified men who fought to escape as best they could. Outside on the frozen fields, a foot deep in snow, Elphinstone had neither the stamina nor the foresight to push on anywhere near to Khoord Kabul Pass. He drew the column to a halt only five miles from Kabul, and it wasn't until two in the morning that the last of the stragglers managed to drag themselves into camp.

That night, a host of 16,000 people huddled in the deep snow, most of them without food, shelter or proper clothing, while in the distance they watched the glow of the flames from the cantonment, which the Ghazis had burnt to the ground. Hundreds failed to rise the following morning, having died overnight of exposure to the sub-freezing temperatures. On the second day's march, the brigades set out in reverse order, with the 44th covering the rear. This was the column's most vulnerable position, as the tribesmen's tactics have always been to harass and plunder their enemy's rearguard and to then make a swift break and vanish into the hills. On this stage of the retreat, the army commanders still believed the parties of Afghans shadowing their progress were part of the escort there to provide protection all the way to Jalalabad. That was until the first war cries echoed across the plains bringing crowds of Ghazi fanatics down on the troops, hacking their way through those who got separated from the main body. Men and guns were overrun in a melee that spread to the centre, where many hapless camp followers were also cut down. Many of those who escaped the tribesmen's daggers met death from a sheer lack of strength to carry on. 'Numbers of infortunates have dropped, benumbed with cold, to be massacred by the enemy', wrote Lady Sale that day. 'Yet so bigoted are our rulers that we are still told the *sirdars* are faithful, that Akbar Khan is our friend!'[37]

The army had marched out of Kabul carrying only five and a half day's rations to take them to Jalalabad. They had no forage for the cattle, or hope of procuring any on the road. There were precious few tents for the men or their families, who were paralysed with cold. It had quite literally become a matter of life and death for the column to make it through the Khoord Kabul Pass, on the other side of which lay ground free of snow, with fodder for the cavalry horses and baggage animals. But they had only covered ten miles in two days' desperate marching, with nearly that distance again still ahead. How many were to perish in the snow or under a Ghazi's knife would depend on how swiftly they reached the village of Tenzin beyond the pass, or so they believed. At this stage the army's greatest obstacle was the terrified horde of camp followers who moved as a chaotic multitude. They pushed ahead, mingling with the troops who found it nearly impossible to march in an orderly fashion, much less form up to repel enemy attacks.

On the third day of the retreat, the column entered the five-mile-long Khoord Kabul, a deep, sunless gorge shut in on either side by a line of high cliffs. Lieutenant Eyre, who stood in awe contemplating the entrance to the pass, wrote, 'The idea of threading the stupendous pass before us, in the face of an armed tribe of bloodthirsty barbarians, with such a dense irregular multitude, was frightful.'[38] As the first contingent advanced the camp followers, sensing the impending danger, broke into a run as one, monstrous and unmanageable mass seeking protection with the soldiers. At that instant the enemy, from their position on the rocks above the canyon, began to rain down a deadly fire on the defenceless knot of people below. The women who rode in front spurred on their horses. The stream snaked between the narrow cliffs, making it necessary for the women to gallop back and forth across the shallow water nearly 30 times. Providentially, they all made it through the gauntlet unharmed, except for Lady Sale, who took a bullet in the arm. With characteristic sangfroid, her diary entry for that day merely mentions in passing that 'I had, fortunately, only one ball in my arm.'[39] Her real concern was for her son-in-law, the army engineer John Sturt, who lay mortally wounded after riding back to the mouth of the pass to search for a friend. She saw to it that Sturt was given a proper Christian burial, the only member of the force to receive this privilege.

That day Akbar Khan, the Afghan conspirator with the most reason to despise the British, made his appearance. Dost Mohammed's son rode up to Elphinstone and Pottinger offering to protect the column in exchange for payment of 15,000 rupees, and an agreement that Pottinger and Captains Lawrence and Mackenzie, the two officers who had been present at Macnaghten's murder, be given over as hostages for General Sale's evacuation of Jalalabad. These disgraceful demands were accepted and the three officers were taken away in captivity. As they rode off, Pottinger later recounted that he heard Akbar Khan shout in Persian to the Ghazis 'Stop firing!', and then in Pashtun, 'Slay them all!', which the Afghan imagined his prisoners wouldn't understand.

The eyewitness estimates of how many lost their lives that day in the Khoord Kabul range from 3,000 (Lady Sale and Kaye) to an improbable 8,000 (Eyre). By anyone's measurement it was slaughter on a massive scale and the survivors who emerged from the pass no longer bore any resemblance to a fighting force.

The remnants of the army struggled on, fighting every step of the way, until on 12 January they were in sight of their next obstacle, the infamous Jugdulluk Pass, which lay a further gruelling three miles uphill. As they approached the entrance hordes of Ghazis were spotted lining the steep escarpments on both sides, but even worse, the column was halted at a barricade of logs and boulders blocking their way into the pass. The Afghans opened up with their *jezails*, throwing the entire column into a scramble for cover, but to little avail. The troops and camp followers were mowed down or crushed under rocks that came crashing down from the heights. For Kaye, the battle in the Jugdulluk Pass was the final nail in the army's coffin:

> The massacre was something terrible to contemplate. Officers, soldiers and camp followers were stricken down at the foot of the barricade. A few, strong in the energy of desperation, managed to struggle through it. But from that time all hope was at an end. There had ceased to be a British army.[40]

At daybreak the following morning, 13 January, all that was left of the army, a party of some 20 officers and 45 soldiers of the 44th, with a pack of some 300 camp followers jostling their way forward amongst the troops, reached the village of Gandamak, still 35 miles from Jalalabad. The soldiers were down to not more than two rounds each in their pouches but the most senior officer left standing, Major Charles Griffiths, formed up the men in a square on a little hillock, determined to give a good fight to the end. The few survivors of this heroic last stand said they faced odds of a hundred to one, and that most of their men had already sustained wounds. The Afghans had surrounded the little band of troopers and were waving their banners and beating their war drums, when suddenly a messenger broke out from the mass shouting to an astonished Major Griffiths that Akbar Khan awaited him in conference. Hopeful of negotiating a surrender deal, Griffiths rode off to the Afghan camp. No sooner was he out of sight, the Ghazis rushed the small group of men left holding the mound. Only Captain Thomas Souter, along with three or four privates who had fallen wounded, were spared. The Afghans never showed any mercy to British officers, but in this curious case Souter had wrapped the regimental colours round him, leading the Afghans to take him for a person of distinction whose life could be exchanged for a sumptuous ransom. The fortunate British survivors were handed over to Akbar, who sent them under escort to a fort in the Lughman Valley, a month's journey away. The 44th of Foot, like the other British and Indian Army regiments that two and a half years ago had marched in triumph across Afghanistan, was no more.

CHAPTER 4

Vengeance is Mine, Sayeth Lord Ellenborough

Lady Florentia Sale was hastily sorting out her personal belongings on the morning of the army's retreat from Kabul. She was looking forward to her reunion with her husband Sir Robert Sale, who was holding out valiantly at Jalalabad in the face of an overwhelming swarm of Afghan insurgents harassing the garrison. Amongst the books to be left behind, she found a collection of poems by the Scottish Romanticist Thomas Campbell. The volume happened to be opened to 'Hohenlinden', one verse of which haunted her day and night on the army's march to annihilation:

> Few, few shall part where many meet,
> The snow shall be their winding sheet,
> And every turf beneath their feet
> Shall be a soldier's sepulchre.

'I am far from being a believer in presentiments', she wrote on the day of departure, 'but this verse is never absent from my thoughts. Heaven forbid that our fears should be realised!'[1] Little did Lady Sale suspect the extent to which those forebodings were to be fulfilled. A more accurate rendition of Campbell's poem would have substituted 'One' in place of 'Few', for on the afternoon of 13 January, a solitary soldier on a dying pony was spotted limping his way towards the Jalalabad fort. Brigadier William Dennie, who was attached to Sale's brigade, was one who did believe in presentiments. Almost from the first, Dennie had boded ill of the force left in Kabul. He would often depress the men's spirits in the mess with his conviction that the Kabul garrison would be destroyed to a man. Reverend Gleig, chaplain to the 13th, remembered Dennie's prophetic words: 'You'll see, not a soul will escape from Kabul except one man, and he will come to tell us that the rest are destroyed.'[2]

Slowly the rider approached the walls of Jalalabad, whilst the men on the ramparts felt their blood curdle as they watched his spectral advance. An escort of cavalry was sent out to meet the soldier in the tattered uniform, who was brought in bleeding and feeling faint and covered with wounds. He was Dr William Brydon, later immortalized by Lady Butler in her painting *The Remnants of an Army*. Brydon was a 31-year-old Scottish surgeon with the 5th Native Infantry, one of the units that had remained behind in Kabul after the departure of the main body of the Army of the Indus. Brydon and five other British officers had managed to escape the carnage of the retreat and had halted at Fatehabad, a village four miles short of Jalalabad. There they were given refreshment by the villagers, who without warning turned on the Feringhees and knifed them, all but Brydon, who achieved a second almost miraculous escape from death. He leapt onto his exhausted pony and spurred it off towards Jalalabad, taking wounds to the knee and to the left hand as he fled. He had received a near-fatal blow to the head from an Afghan knife, but was saved by a copy of *Blackwood's Magazine* rolled up under his forage cap. It was the wound to the knee that ironically saved his life: when his Afghan assailant saw him stoop forward, he thought Brydon was about to draw a pistol, so he turned and galloped off.

That night lanterns were suspended from poles at different parts about the fort's ramparts. From time to time the buglers sounded the advance, in the hope that one or another of the beacons might guide some wanderer to safety. This was repeated for several days and nights but not one man, British or native, came in.

Unbeknown to Sale and his fellow officers, Brydon was not

the only European left alive from the column. The brigadier would have been overjoyed to learn that his wife Florentia had been one of the handful of prisoners, who included officers, 12 women and 22 children, who were taken hostage by Akbar Khan. To be sure, this was nothing to do with some benign act of compassion. Akbar was simply hedging his bets, knowing that if the military situation should turn against him, he had in his hands the perfect tool to bargain his way to safety, though in the end it served him to no avail.

Akbar had cunningly persuaded Elphinstone to give up the women and children, alleging that while he was able to restrain his fellow renegades, he had no means to stop the Ghazis indulging their orgy of slaughter. This was discussed with Elphinstone and Shelton in conference at Akbar's camp under a flag of truce. Whatever his physical or mental failings, the general was a decent man who would have gone to any lengths to rescue the women and children from being massacred. Directly he agreed to Akbar's offer, for good measure Elphinstone, as well as Shelton and Pottinger, were told that they, too, were to accompany the prisoners. Eyre, who was amongst the fortunate hostages, wrote that no one at the time had much confidence in Akbar's integrity. 'It seemed as though we were but too probably rushing from a state of *comparative safety* into the very jaws of destruction.'[3]

The prisoners bade a hasty farewell to their friends, whom they were never to see again, and on 9 January they mournfully followed their captors to a spot about two miles distant from the killing fields. Thus began a nine-month odyssey of suffering, tempered by brief periods of unimagined luxury – wood fires, greasy pilau rice and *poshteen* (sheepskin cloaks) for protection against the cold – through the treacherous mountain passes of Afghanistan.

As the little band of hostages was led off into the wilderness several days later, Brydon was recounting to Sale and the staff officers his adventures with the retiring army and his remarkable escape. All discipline had come to an end at Jugdulluk, where the column was disintegrated into confusion. In the midst of this free-for-all, in which soldiers and camp followers fused into one terrified mass, Brydon came across a saddler of Shuja's cavalry, who knew that he was mortally wounded and begged him to take his pony to prevent it falling into enemy hands. No sooner had he uttered these words, the man fell dead at Brydon's feet. This was the pony on which Brydon made good his flight to Jalalabad. There is a tale sometimes heard in officers' messes that the heroic pony's remains were sent back to Britain for burial, but this quaint little anecdote has never been verified. Brydon made his lucky escape from Futehabad only to run into some 20 men drawn up in the road, unarmed villagers who began picking up stones and hurling them at the Feringhee. Gripping the reins in his teeth, Brydon put his pony into a gallop and cut right and left with his sword as he went through his attackers. Almost immediately, he was met by another group of tribesmen, one of whom fired his *jezail* at Brydon, shattering his sword about six inches from the handle. The ball lodged in the poor pony's loins, forcing Brydon to prick the animal with his broken sword to spur him on. At Jalalabad, Brydon was taken to the Sappers' mess, where his wounds were dressed and, after a good dinner, he told his astonished listeners that apart from the knife cut to his head, he had no recollection of how and when he took the wounds to his knee and hand, or the musket ball that grazed his leg.

The repercussions of the British Empire's unprecedented military debacle at the hands of an Afghan rabble armed with daggers and *jezails* swept like a shockwave across the world. To the Muslims not only of Afghanistan, but of India, Turkey and elsewhere, this was greeted as an unequivocal sign of vulnerability. At last, the arrogant Feringhees who claimed a right to dominate their lands had been routed. *The Times* correspondent in Constantinople reported that the disaster in Afghanistan had already begun to produce a negative impact on British influence in the Muslim world. 'Since the last overland intelligence', the dispatch said, 'it has been asserted on more than one occasion by Turkish authorities that the late reverses of England had suddenly reduced her to the place of a third-rate power.'[4]

When Lord Auckland was told the extent of the army's disaster, he was reported to have thrown himself on the ground in uncontrollable sobs of despair. Auckland suddenly realized that he had been disastrously misled by his trusted envoy at Kabul. Even in late October, scant days before the uprising, Macnaghten continued to report with glowing confidence the increasing tranquillity of Afghanistan. He assured Auckland that the insurrection of the hill tribes seemed to have been suppressed and there was nothing stirring in the neighbourhood of Kabul to create anxiety or alarm. The governor general later willingly accepted the blame for not having foreseen the catastrophe – how could it have been otherwise? – and he confided as much to Hobhouse of the Board of Control, albeit with a half-hearted attempt at self-exculpation. He reminded Hobhouse that Macnaghten, Elphinstone, Burnes and the others had repeatedly assured him that the situation at Kabul was improving. True, yet had Auckland taken the trouble to seek an opinion from the military commanders in the field, rather than relying on reports from the 'spin doctors' of the day, he might have drawn a more sobering conclusion on the reality of Britain's position in that country.

Auckland, a lifelong Whig, had sent in his resignation when Peel's Conservative Government came to power in July 1841. Kaye relates that Auckland had expected to embark for England 'a happy man and a successful ruler' but when the awful truth came out, he overnight turned into 'the most luckless of rulers and the most miserable of men'. The scales had fallen from the eyes of the governor general, 'only to show him the utter hopelessness of the case. In this terrible emergency, he seems to have perceived, for the first time, the madness of posting a detached force in a foreign country, hundreds of miles from our own frontier, cut off from all support by rugged mountains and impenetrable defiles.'[5] Auckland would now be going home in disgrace, the man who presided over the destruction of an army. All his grand plans had broken down in the massacre 'of which history has few parallels', Auckland himself wrote, adding:

I look upon our affairs in Afghanistan as irretrievable ... For there is yet many a point of deep and painful anxiety in regard to the insulated posts in which detachments of the Army are placed, and I fear that we are doomed to hear of more horror and disaster.[6]

For having sanctioned an invasion that cost the lives of 16,000 European and Indian troops and civilians, with the aim of imposing an unwanted ruler on a country that was at best of questionable relevance to the security of British India, Auckland was rewarded with an appointment as First Lord of the Admiralty – a job, it must be acknowledged, he exercised with some distinction. Even as Auckland disembarked from the frigate that had carried him home to London and to the Eden family's house in fashionable Kensington, the last threads of his Afghan misadventure were coming unravelled.

One morning after Burnes's murder, a delegation of rebel leaders climbed the hill to the Bala Hissar to seek an audience with Shuja. The purpose of this visit, quite to the Saddozai chief's own astonishment, was to ask him to stay on as emir of Afghanistan once the British had departed. The only stipulation was that he agreed to give away his daughters in marriage to the chief conspirators, thereby cementing the union between the two contending Pashtun factions. Shuja refused to give in to this demand, however he did gain the rebels' favour by agreeing to appoint one of their chiefs, Zaman Khan, as his vizier, or chief minister.

Akbar later told Shuja in conference that his people looked to the emir for some gesture of loyalty to the insurgents' cause. This meeting took place in April 1842, three months after the army's evacuation of Kabul. Shuja believed that Akbar had the British forces on the run everywhere in his kingdom so, always keen to back a winner, the emir consented to muster his troops and artillery to join Akbar in besieging Sale's garrison at Jalalabad. On 5 April, Shuja rode out of the Bala Hissar on his white charger, accompanied by a small escort, to set off on the march with the royal regiment encamped nearby. A few hundred yards outside the citadel, the emir was gunned down by his faithful vizier's son, Shuja ud Dola. Shuja's body was cast into a ditch and a few days later, a multitude of widows began trundling their way back to India. One historian somewhat illiberally sums up Shuja's life as that of a perennial failure, who 'betrayed his benefactors, resisted his liberators, and died by the hand of an assassin – for an Afghan he could have no more glorious close to such a career'.[7]

One man who did not look upon Britain's affairs in Afghanistan as 'irretrievable', nor anticipated more 'horrors and disasters' in that country, was Auckland's successor Edward Law, afterwards Lord Ellenborough. Like Hobhouse, Ellenborough had served as president of the Board of Control, four times in fact, and he had only held office for a month in his last term when Peel offered him the Governorship General of India, with a specific brief to clear up Britain's position in Afghanistan. Ellenborough gladly accepted the post and on 28 February 1842, his East Indiaman docked at Calcutta.

Ellenborough was in almost all respects a different man to Auckland. A powerful orator, twice married, a genuine authority on military matters, he was always attracted to army life and he was also in every sense of the word, his own man. With Ellenborough in charge in Calcutta, there was to be no scheming, self-serving Macnaghten in Kabul, Wade in Ludhiana or Ellis in Tehran – or Palmerston in London, for that matter – spreading dangerous panic over a Russian threat to India's borders. On the other hand, there would be little need for such goading, for Ellenborough had always maintained a bellicose attention to Indian politics, more particularly in relation to possible Russian aggression. The new governor general had arrived in India determined to repair the damage that had been inflicted by his predecessor and, most importantly, to restore British honour and prestige. When that job was finished, he was destined to return to England to succeed Auckland as First Lord of the Admiralty, a post that was becoming something of a dumping ground for former Governors General. An obsession with honour was a salient trait of Ellenborough's character. In 1830 he divorced his second wife after discovering her adulterous affair with the German Prince von Schwartzenberg. But divesting himself of his unfaithful wife did not salve Ellenborough's injured pride: he wanted revenge, so he challenged the nobleman to a duel, receiving £25,000 damages.[8]

On the day that the unloved Shuja fell to an assassin's bullet, Akbar Khan and his men were massed in strength close to the Jalalabad ramparts. Sale had been holding out with 1,500 men against a force of 5,000 Afghans who, apart from their numerical superiority, enjoyed ready access to food, animal fodder and ammunition. Then, on 19 February, Jalalabad was rocked by a severe earthquake which undid in one hour all that it had taken the garrison three months to accomplish. This was Akbar's cue to lead his infantry and cavalry through the walls that had been torn asunder by the quake and destroy the garrison. But he hadn't reckoned with the army engineers' agility in repairing the damage, the men working day and night to rebuild the fort's defences so that within less than a fortnight the parapets were entirely restored, every battery was back in place and the ramparts doubled in thickness. Akbar had hesitated, fatally so, when he should have closed in quickly on Sale's exposed position.

By failing to attack in the earthquake's immediate aftermath, perhaps fearing an aftershock or simply ignorant of the extent of the damage to the fort, the Afghan chief's doom was sealed. Sale too needed to move with alacrity. It had now become a daunting and highly dangerous task to send out foraging parties to gather in cattle and sheep, with Akbar's men camped only a few hundred yards from the perimeter walls. A shortage of ammunition had also been an acute problem since taking possession of the fort. Some of the soldiers had devised an ingenious scheme to address this dilemma. They carved a block of wood into a human figure and dressed it in Sale's spare uniform. When the mannequin was shown over the ramparts brandishing a sword, it attracted a hail of bullets, which the men picked out of the parapet wall for reuse. This device worked for only a while, however, before the attackers caught on to the ruse. At the beginning of April, Akbar moved his troops closer to the town in preparation for what was expected to be an all-out assault on the garrison. The Afghans had ringed the walls with sangars (stone breastworks) and there were rumours of the enemy starting to tunnel under the walls to place explosives.

Sale took action on 7 April by ordering an attack, one that ranks as amongst the most glorious feats of arms of the Raj. It was an all-or-nothing gamble with three columns comprising almost all the garrison's fighting men, whilst keeping only a skeleton force behind to defend the fort and look after the sick and wounded. Three of Sale's most distinguished senior officers were chosen to lead the sortie. Lieutenant Colonel Thomas Monteith took 500 men of the Bengal Native Infantry on the left, while the centre was made up of another 500 troops under Colonel Dennie, with Captain Henry Havelock on the right flank at the head of 360 British and native infantry. The Light Field Battery was brought up under Captain Augustus Abbott, along with a small troop of cavalry. Sale's strategy was to take Akbar by surprise. He assumed the last thing the Afghan would anticipate was for the besieged garrison to go on the offensive – and he was proved right. The infantry assembled before dawn at the west gate facing the Kabul River and the artillery and cavalry was drawn up at the south gate, with orders to march direct upon Akbar's camp, to burn it to the ground and to capture his guns.

In the tradition of more than one great British victory, the assault came within a whisker of ending in disaster. Between two and three hundred of the enemy occupied a ruinous old fort half a mile from the west gate. Dennie's column led the advance and immediately came under heavy fire from this outpost, but instead of pressing on at all speed across open ground, Sale halted the column until the guns could be brought up to give the fort a pounding. One chronicler condemned this as a serious tactical blunder that cost the lives of Dennie and several of his men, and delayed the advance of Monteith's column on the left, leaving Havelock's troops exposed on the right flank:

The assault was a complete failure. The whole mass of the Afghan horse swept down upon Havelock. Coolly forming square, Havelock repulsed them with heavy loss (on the enemy's side) and Sale, presently awaking to his folly, abandoned the assault upon the fort and resumed the original plan of the advance.[9]

The troops pressed on, ignoring Akbar's cannons, some of which had been taken from Elphinstone. They charged the Afghan positions with musket fire and bayonets, rushing the enemy positions with fury, in the awareness that retreat would spell doom. Akbar's soldiers broke and fled, many being swallowed up by the river, which in their panic they attempted to cross. By nightfall there was not an Afghan within eight miles of the garrison. The siege of Jalalabad, for all intents and purposes, had come to an end.

When Elphinstone's request to be relieved of his command reached Calcutta, the Government appointed General George Pollock to replace him as commander-in-chief in Afghanistan. Pollock was a rarity amongst the leading figures of the First Afghan War, in that he was not a Scotsman, but a dyed-in-the-wool Londoner, born at Piccadilly in the heart of the capital. He was not regarded as a brilliant commander, but on the other hand brilliance was not the quality Ellenborough and Sir Jasper Nicolls were after in the man dispatched to relieve Sale's brigade at Jalalabad. Ruthless determination was needed and there was no shortage of that in Pollock's nature, a soldier who had fought his first battle in India at the age of 19. The general was faced with the formidable challenge of being the first commander in history to fight his way through the Khyber Pass, a task the pious, imperturbable Pollock was to accomplish with distinction, where military legends like Alexander, Tamerlane and Babur had met with failure.

One of Auckland's last edicts as governor general was to assure Pollock a force of at least 10,000 men, to be assembled at Peshawar at the mouth of the Khyber Pass for the march on Afghanistan. Auckland was so severely shaken by the Afghan disaster that he had summoned regiments as far afield as Burma and Madras, and he also begged for three more of the Queen's regiments to be sent from England to join the task force. Nicolls kept his head in this wild scramble for troops, realizing that these regiments might well be needed elsewhere to crush the threat of popular revolt once India's Muslim population got word of the annihilation of the Army of the Indus. The Government had in the past been obliged to put down outbreaks of rebellion by peasants, tribal communities and princely states, such as the Kol Uprising of 1831 and the Kutch Rebellion, which lasted from 1816 until 1832. Disaffection in the sepoy ranks was likewise not unknown in this period, hence British authority in the years preceding the great 1857 Mutiny could by no means be taken for granted.

Auckland dithered, but not so his successor. With Ellenborough in Calcutta, there was no doubt in anyone's mind that the reins of power had passed to a far more vigorous man of action. Norris likens Ellenborough's arrival at Government House to that of 'a *nouveau riche* owner surveying a crumbling family mansion before the family had moved out. His brisk and proprietorial orders and his haughty manner with officers and servants alike assaulted the ears of the Edens . . .'[10] Whatever offence the new governor general's rude mannerisms may have caused Auckland's sensibilities, Ellenborough was the man of the hour, with a firm grasp of military matters and an excellent relationship with Pollock, who was to command this risky second invasion of Afghanistan: Britain could not afford to sustain another reverse in that country. In carrying forward his task, Ellenborough could count on the support of people who mattered – the Duke of Wellington, Peel and Nicolls[11] to name but a few – all of whom were anxious to see a speedy restoration of Britain's prestige in the world.

No sooner had Ellenborough landed in Calcutta, he dispatched a letter of support to Pollock, expressing his 'approval of a good soldier'.[12] Pollock was given command of the invasion force to be known as the Army of Retribution, as grandiloquently named as its predecessor the Army of the Indus, and a fore-echo of the more contemporary Operation Enduring Freedom.

The tide began to turn on 5 February 1842, when Pollock arrived at Peshawar to lead the Army of Retribution to the relief of Sale's brigade at Jalalabad, which still had two months of siege and an earthquake ahead of it. What the general found in the Sikh frontier outpost did not fill him with much cheer: a disheartened, defeated garrison prostrate with malaria and dysentery. Pollock learnt to his dismay that Brigadier Charles Frederick Wild, who had arrived ahead of Pollock, acting in response to Sale's urgent calls for help, had taken his four native infantry regiments with an attachment of gunners into the Khyber Pass in an attempt to break through to Jalalabad. The Swiss-born Wild's fatal mistake was to split his force, moving out two regiments at midnight on 15 January, while the brigadier himself planned to bring up the rest of the force, with baggage, cattle and provisions, the following day. No sooner had Wild and his Sikh regiments moved into the pass than his men were savagely beaten back to the entrance of the Khyber. The Sikhs took to their heels, spreading confusion amongst the forward contingent, and with this almost the entire force fell back on Peshawar in disarray. Wild rallied those who had stood their ground for another assault, but his troops were cut to pieces at a barricade the Afridi tribesmen had thrown across the narrowest spot on the road, spanning some ten feet in width. Back in Peshawar, it took all of the newly-arrived Pollock's unassuming leadership to raise the morale of nearly 2,000 ill troops, whom he personally visited in hospital, while maintaining the spirits of the now inactive force he had brought up from the Punjab, and quelling the first stirrings of mutiny by Wild's native battalions, who made it known they had no intention of joining a second attack. This frustrating state of affairs carried on until late March, when Pollock received his last reinforcements from British India, and was finally prepared to start his advance on the Khyber Pass.

The army had been sitting idle in Peshawar for nearly two months and apart from the demoralization this caused amongst the troops, it was also a drain on the Government of India's finances, which were under close scrutiny in London. Fortunately, Pollock found a benefactor in the Italian General Paolo Bartolomeo di Avitabile, another of those great European eccentrics of the day in the tradition of Masson and Burnes, who journeyed to far-flung places to seek their fortunes. Avitabile, who had served as one of the 'white officers' in Ranjit Singh's army, advanced large sums of money to Pollock's field treasury, thus relieving the strain on the army's coffers. Avitabile was appointed Governor of Peshawar in 1834, the year after the city had been taken by Ranjit Singh's forces. Avitabile's iron-fist rule made a place for him in local folklore and even today naughty children in Peshawar are brought to control by invoking the name of *Abu Tabela*, a Pashtun corruption of 'Avitabile'. Visitors arriving at Peshawar's eastern gate from the Punjab would be greeted by half a dozen gibbets, to which he would add two or three on festival days. One horrified European visitor recorded that Avitabile would in the morning hang a dozen unhappy culprits, look to the payment of his troops, inspect his poultry yards, set a-going a number of musical boxes and attend to other domestic affairs 'all before dinner at noon'. Moreover, a large treble gibbet stood at every corner of the city, each of which with 17 or 18 malefactors hanging on it, according to Lieutenant John Greenwood, who served with the Army of Retribution. He says this was meant as 'a gentle hint to the inhabitants to be on their best behaviour. I believe there was very little ceremony made with them. If a man looked sulky, he was strung up at once, in case he should be disaffected.'[13]

Avitabile's methods were intended to instil terror, and even the British troops passing through Peshawar, who could by no means be described as squeamish, were awed by this display of brutality. Lieutenant James Slater Cumming, a young officer with Pollock's force, gave a vivid description of what he saw when marching into the city:

> On the right of the road, issuing from the town, stood two gallows on a high mound, from one of which hung three criminals, from the other twelve criminals. Some were in the last

stage of decomposition, exhibiting their fleshless bones dangling in the air. Others had apparently been but a short time hanging ... I fear that he [Avitabile] is a ruthless man, and holds the life of a human being and a brute in equal estimation.[14]

Avitabile succeeded in building a fortune and taking it home with him. Thanks to the loans made to the British, Avitabile was able to transmit to the Bank of England a considerable fortune, which he later reclaimed on his return to Europe. He went back to Italy, where he built a grand home near Naples, and died under suspicious circumstances soon after marrying a local girl. The ensuing legal battle over his inheritance, and the many distant relatives asserting their claims, has made 'Avitabile's cousin' a byword in Italy.

Pollock began his advance on the Khyber Pass on 5 April, leading a massive force of 12,000 men, though it could hardly be said the troops moved out in high fighting spirits. 'The Sikhs', according to Fortescue, 'were still terrified at the prospect of entering the defile. Avitabile averred that Pollock was marching to certain destruction. The sepoys of Wild's brigade were still deserting.'[15] There was indeed much to shake the nerve of a commander about to embark on an historical mission. At 3 a.m., without call of bugle or beat of drum, two companies of sepoys of the 31st were sent into the pass as skirmishers to clear the heights on either side, with two nine-pounder guns placed at the mouth. The main body of the force that followed was a formidable army of 13 companies of native and British troops, sappers and miners, five squadrons of mounted infantry, the 9th Queen's Regiment, the 10th Light Cavalry, nine pieces of artillery and many hundreds of pack animals and camp followers. The heights were swarming with Afridi tribesmen, who to the army's good fortune realized they were outgunned and fell back at the unexpected sight of sepoys rushing the hills with fixed bayonets. Having cleared both flanks, the main column then made its way through and pursued its march with little opposition to Ali Masjid, a spot five miles within the pass at its narrowest point.[16] There are two small forts here that stand on the summit of an isolated rock and this has always been regarded as the key to the Khyber Pass. These positions had been evacuated in haste by the enemy upon sighting the approaching multitude of men and guns. At the break of dawn, Lieutenant Greenwood recalled finding many mementos of Wild's disastrous attempt at forcing the pass. 'Dead camels and horses, broken and defaced military accoutrements and human bones lay scattered about in every direction.'[17]

Once past Ali Masjid, Pollock arrived at a narrow defile about half a mile in length, with perpendicular cliffs on each side. It took nearly the whole of the day to get the guns through this cleft and, as Greenwood reflected, it is a wonder they ever got through it at all. Looking up at the cliffs, he saw that 'if only rocks were thrown down by the defenders posted on the top of them, every individual below must be crushed'.[18] Greenwood was jammed by the press of baggage animals for six hours, the time it took the column to march single file past this narrow spot. 'I looked up at the threatening heights above us, and felt how utterly helpless we were. Six men from above could have annihilated the whole of us.'[19] In two days' time and having taken only a handful of casualties, Pollock's army emerged from the western end of the Khyber Pass to set a milestone in military history.

Sale's garrison should have been jubilant at the news of Pollock's breakthrough, but the mood at Jalalabad was instead one of despondency. Afghan spies were putting about stories that Pollock's force had been repulsed in the pass, reports that Akbar, to rub salt in the wound, celebrated with a victory salute from his guns. That was on the morning of 6 April, just when Pollock was climbing almost unopposed towards the 3,600-foot summit of the pass. There were a few uncomfortable hours in the garrison when several senior officers, who had proposed most vigorously sending out a column to attack Akbar's positions, contemplated disobeying Sale's orders, which were to hold fast. Providentially, it never came to blows, as later that day Sale withdrew his opposition to the attack, which as we have seen took place on 7 April and ended in an Afghan rout. A few days later the happy truth came in that Pollock had forced the Khyber Pass after all, and by 15 April the Army of Retribution stood assembled on Afghan soil, preparing to march to the relief of Sale's brigade, which no longer needed to be rescued. Pollock's force was now camped within seven miles of Jalalabad. Then on the 16th the band of Sale's 13th Light Infantry[20] went forth across the fields that had been cleared of Afghans to join their comrades in a jubilant reunion. 'There was a hearty cheer on both sides', recalled Greenwood, 'after which the musicians facing about began, according to immemorial usage, to play the strangers in.'[21] As they neared the fort, the bank struck up the Jacobite melody 'Oh but ye've been lang o' coming.'

Military victory in Afghanistan is an ephemeral business. It comes with a stinger in the tail, and it is the 'defeated' Afghans who more often than not have the last laugh. Sale's 'Illustrious Garrison', as it came to be known in Britain, and the army that had heroically battled its way from India to rescue it, began to languish. The rearmost of Pollock's regiments had gathered about Jalalabad in the first week of May, in the worst of the hot season that brings in its wake dysentery and heat exhaustion. The troops were encamped on a sandy plain with no shelter from the punishing sun that beat down, with temperatures soaring to 120°F. Provisions were hard to come by and what little water was available was insalubrious, when not outright lethal. Day after day, the death toll rose from dysentery and fever. The troops had to suffer as well the stench of the putrefying carcasses of unburied horses and camels. Pollock awaited in vain a signal from Government House telling him what course of action to pursue. The Government in London was naturally delighted with the successful relief of Jalalabad, but there was a marked lack of enthusiasm for taking the campaign further and risking the loss of a second army.

It was a frustrating position for Ellenborough, for British honour had not been restored as he would have wished. Pollock had tried and failed to negotiate the release of the British hostages held by Akbar Khan. This was hardly surprising, given that the prisoners represented the Afghan's warlord ultimate bargaining tool. Then in June Ellenborough took matters in hand and issued instructions of a morally ambiguous nature, leaving Pollock to interpret the orders as he saw fit, and presumably bear responsibility for the outcome. The general was to use his discretion to take the army forward or, if he preferred, to beat a retreat to India. Pollock did not hesitate for a moment: he had not led more than 10,000 men across the Khyber Pass into Afghanistan only to turn them around and take them home, having barely engaged the enemy. Pollock's brief was to rescue the prisoners in Akbar's hands. While Ellenborough had deliberately omitted any reference to Kabul, by a fortuitous coincidence, the General must have mused with a smile, the army had by necessity to march in that direction to reach the hostages. There remained the task of teaching the Afghans a proper lesson, and of course locating and freeing the hostages. So on 6 August, Pollock directed Sale to advance eastward to the town of Futtehabad, which lies on the road to Kabul.

On 21 August, the residents of Fatehabad emerged from their homes to witness a gathering of humanity and animals of apocalyptic proportions. A division of more than 10,000 British and native infantry regiments had marched from Jalalabad, with batteries of guns and cavalry, accompanied by 5,000 Sikh soldiers and no fewer than 40,000 camp followers, plus many thousands of camels, horses and bullocks. The force carried on to Gandamak, while rumours of the hostages' fate were passed along the line. It was asserted, correctly so, that the prisoners had been taken some 100 miles west of Kabul to Bamiyan. Akbar was threatening to drive them all beyond Afghanistan's borders to Bokhara, where they would be sold as slaves unless the Government was prepared to launch yet another invasion beyond British India's frontiers. The handful of officers who had been left behind in Kabul, too ill to cope with the rigours of the journey, were later brought up to join their fellow hostages. The exception was General Elphinstone, who had died on 23 April on the road from Tenzin to Kabul. Captain George Lawrence was with Elphinstone until the end. He related that in his final moments, the general muttered in a failing voice that his greatest regret was not having died with his men on the retreat. There was decency in the heart of that soldier, who had thrust upon him a duty he was never capable of discharging, and in which he found little sympathy in his fellow officers.

And what of Nott in Kandahar? The irascible old warrior, like Sale, had ignored Macnaghten's orders to evacuate his position and, as events were to prove, it was unquestionably the right decision. Nott was in command of a strong force of more than 6,000 men and 22 guns, and though short of cavalry he was able to make several successful sorties against the Afghan besiegers. Moreover, General Sir Richard England in Quetta had responded to a call to further strengthen the garrison by setting off with a brigade towards Kandahar. By July 1842, Nott's men were eager to wind up the campaign and go home, after several months of almost uninterrupted skirmishing with Afghan raiders. There hadn't been much action since May, when Colonel George Wymer was sent out with a large column to relieve the garrison of Kalat-i-Ghilzai. A mixed brigade was dispatched in support of Wymer's force holding out at that outpost, but the brave little garrison had already successfully repulsed a most determined assault six days before the relief force arrived. Major Alexander Hamilton notes:

> It was here that the Shah's 3rd Infantry, which formed two-thirds of the tiny garrison, earned the title of 'Kalat-i-Ghilzai Regiment' and a place in the regular army as the 12th Native Infantry. Reliever and relieved not being strong enough to remain at Kalat-i-Ghilzai were withdrawn to Kandahar.[22]

Nott was by now desperate to take his troops to Kabul, having little left to do in Kandahar. Striking a strong blow against the capital and the Afghan rebels in occupation of the Bala Hissar would compensate for almost three years of semi-exile, far from where the main action was taking place. But Ellenborough's hands were tied by London. The governor general would have gladly turned the army's most aggressive brigadier loose on the Afghan insurgents holding Kabul, however a more tactful approach was needed to avoid provoking a confrontation with Peel's pusillanimous Government.

At home, the press had already loosed a bombardment on Whitehall for having sanctioned the first Afghan invasion, and the newspaper assault was to roll on and gather momentum well after the Army of Retribution's withdrawal from Afghanistan. For instance, this particularly ironic attack published in *The Times* took the wind out of the Government's sails a few weeks after they were congratulating themselves on a job well done and handing out honours to the military commanders: 'This nation spent £15 million on a worse than profitable effort after self-aggrandisement in Afghanistan, and spends £30,000 a year on a system of education satisfactory to nobody.'[23]

Nott moved from Kandahar on 8 August and by the end of the month he was two days' march from Ghazni. Much to Ellenborough's chagrin, the citadel over which the Army of the Indus had so heroically raised the Union Flag after its hard-fought battle of July 1839 had been recaptured by the Afghan insurgents the very day the new governor general landed in Calcutta. Ellenborough and every officer in the army were even more outraged over reports received of the brutal treatment that had been inflicted on the captured British officers, who were forced to surrender the fort when their water supply ran out. The Afghans behaved with particular barbarism towards the sepoys that were taken prisoner, selling some into slavery and murdering the rest outright. Nott exacted his revenge with an equal lack of niceties: on 5 September he moved on Ghazni and directed the Bengal engineers to reconnoitre the works, under escort of the 16th Native Infantry and a party of irregular cavalry. It took less than an hour of steady bombardment for the general's troops to drive the enemy from the lower slopes around the fort, followed with an onslaught by the brigade's nine-pounder guns. Three days later, he sent a dispatch to Ellenborough informing him that:

> The engineer officers, sappers and miners, and infantry working parties were employed ... during the night of the 5th, in erecting a battery for four 18-pounders. These guns were moved from camp before daylight on the morning of the 6th, but before they had reached the position assigned them, it was ascertained that the enemy had abandoned the fortress.[24]

The Afghans rank amongst the world's most formidable hill fighters, but in open combat they are no match for a disciplined European officered army. Not only had the defending garrison been routed, on taking possession of the city the engineers were given two days to blow it up. In Nott's own words, 'I directed the city of Ghazni, with its citadel and the whole of its works, to be destroyed.'[25] Lieutenant James Rattray, famous for his Afghan lithographs, noted on one he did of the fortress when he last glimpsed it, 'Far-famed Ghazni was a heap of smoking ruins.' The troops were able to recover more than 300 of the sepoys of the 27th Native Infantry who had been sold into slavery and dispersed in villages 30 and 40 miles around Ghazni.

Nott covered the 310 miles from Kandahar to Kabul in less than six weeks, fighting many skirmishes and one general action on the road. The two armies converged on Kabul in mid-September and on the 18th Pollock rode into Nott's camp to greet the hero of Kandahar. The cantankerous Nott, who was as ever nursing a grudge at the Government's preferential treatment of Queen's regiments over the native sepoy units he commanded, was not in the best of spirits that morning. His senior officer had beaten him to Kabul by two days.

The Army of Retribution was living up to its name. Nott had left Ghazni a pile of rubble, Pollock had given the Afghans an unsparing hiding at Tenzeen and the Jugdulluk Pass, where the remnants of the Army of the Indus had been massacred by the Ghazis. In these actions, Pollock routed a force of 16,000 of Akbar Khan's men on 13 September, and three days later he was in Kabul. But before sowing more destruction the task was to extricate the hostages still held perilously in Akbar's grasp. Pollock ordered his military secretary Sir Richard Shakespear to assemble a squadron of 600 Kuzzilbash horsemen and ride like the wind to Bamiyan to find and liberate the prisoners. The orders were issued not a moment too soon, for when Akbar learnt of the fall of Kabul, he informed his prisoners that they would soon be out of reach of their British liberators. Akbar was in league with the ruler of the khanate of Kulum, who had already dispatched 2,000 men to the Frontier to escort the captives to his capital, where they would later be sold as slaves in Bokhara. Nott was later instructed to send a brigade to assist with the rescue mission, but he complained so bitterly about his men being exhausted after their march from Kandahar (he also had little interest in providing Pollock with basking glory) that after a tense encounter in Nott's tent, Pollock agreed to send Sale to Bamiyan in his stead.

Ellenborough's alarm at the hostages' plight and the rancour between Nott and Pollock were all for nothing, for in the end it was not Shakespear or Sale who set the prisoners free, but the ever-resourceful Pottinger, who was himself one of the hundred being held in captivity.[26] At the little band's darkest moment, Pottinger had a quiet talk with Saleh Mohammed, Akbar's deputy in Bamiyan who was in charge of the troops guarding the prisoners. This Afghan was one of the conspirators who had fought the British, but he was also a man of flexible loyalty, who had previously been with Shuja. He knew that Pollock's army was closing in and he had no desire to be caught on the losing side.[27] Without too much effort, Pottinger was able to persuade Saleh Mohammed to release the hostages and even escort them to the British lines, in exchange for a cash payment of 20,000 rupees and a pension of 1,000 rupees a month for life. There is no record of how many monthly payments he lived to enjoy after Akbar's men discovered his betrayal.

The band of captives was settling down in camp on the first night of their journey to Kabul when a horseman rode into their camp with the glad tidings that Shakespear was making his way up to meet them. 'Our joy and thankfulness at the receipt of this intelligence are not to be described', recalled Captain George

Lawrence, 'and little sleep did any of us have for the rest of that night.'[28] They broke into cheers of joy and tearful embraces the next morning when a cloud of dust announced the arrival of Shakespear's column – all but Shelton, that is, who complained that as senior officer he should have been the first to be greeted by Shakespear. Next came Sale, whose wife recorded in her diary on that day, after more than a year's separation from her husband, 'It is impossible to express our feelings on Sale's approach. To my daughter and myself happiness so long delayed as to be almost unexpected was actually painful, and accompanied by a choking sensation, which could not obtain the relief of tears.'[29] For all his tough exterior, Sale too on this occasion was unable to repress his feelings of joy and relief. Captain Colin Mackenzie trotted up alongside the brigadier on the road and remarked, 'General, I congratulate you.' The defender of Jalalabad turned and tried to force a reply, but he was choked up with emotion. 'He made a hideous series of grimaces, dug his spurs into his horse, and galloped off as hard as he could.'[30]

The ragtag column of freed prisoners was barely recognizable when they rode into Kabul. The men wore long beards and heavy moustaches and the women's English complexions were turned a ruddy brown after many months' exposure to the sun. All were dressed in native garb and could easily have been taken for a party of Afghans, had there been one in sight. The streets of Kabul fell into an eerie silence when Shakespear and Sale, riding at the head of the column, entered the city on the return journey from Bamiyan. The shops' shutters were drawn and people stood huddled in doorways, as if in anticipation of an approaching cataclysm. The terrified citizens did not have long to wait for their forebodings to become reality. There was at first some debate between Pollock and his senior officers over whether to lay waste to the Bala Hissar. The great fortress was in the end spared that fate, not so much out of respect for a majestic piece of architecture, but for straightforward strategic considerations. Everyone knew that Dost Mohammed would soon be on his way back from exile in British India. If the Barakzai king was at long last to be recognized as Britain's ally, blowing up his palace might not be interpreted as a gesture of friendship. So the Bala Hissar was off the hook, but not so the Grand Bazaar, where the remains of Macnaghten's corpse had been so wantonly displayed. It did not get off lightly.

On 9 October, Pollock ordered his chief engineer to raze the vast marketplace, after giving timely warning to its inhabitants to vacate their homes. Explosions continued to rock the great souk for two days, by which time not a wall was left standing. Though all efforts were made to save the rest of the city from wholesale devastation, there was little the officers could do to prevent the soldiers and camp followers pouring in to loot and pillage to their hearts' content. Much of this work was done by the Hindu sepoys, who were maddened by the sight of clothes and accoutrements that had belonged to their dead comrades now hanging in the market stalls. Greenwood described the scene in grim detail:

> We continued the work of destruction until night closed upon us and then returned to camp tired enough. Many of our men looked just like chimney sweepers from the fire and smoke. On succeeding days other parties were sent, and the city of Kabul, with the exception of the Bala Hissar and the Kuzzilbash quarter, was utterly destroyed and burned to the ground. The conflagration lasted during the whole time we remained encamped in the vicinity, and we still saw it when entering the Khoord Kabul Pass, on our return.[31]

Retribution did not end with the knocking down of the Kabul bazaar and the raising of the Union Flag over the Bala Hissar. General Pollock dispatched a punitive expedition to take places that remained in the hands of the insurgents. Istalif, some 40 miles north of Kabul in the Kohistan region, was the first to experience what in later Frontier warfare came to be known as 'butcher and bolt' raids, that is, the quick in-and-out demolition of a hostile village or fort, usually involving the burning of crops and seizure of livestock. The army was not playing for low stakes, and so it was in Istalif, where on 29 September General Sir John McCaskill[32] stormed the fortress, sending its defenders fleeing to the hills, after which engagement the eccentric McCaskill was found sitting under a tree munching a basket of Kabul plums. That night, as the column wound its way towards its next objective, the darkness of the Kohistan hills was illuminated by the blazing town.

Charikar lies only eight miles from Istalif and this was next in line for a visit by McCaskill's avenging troops. When the column arrived at the town where the Gurkha brigade had been wiped out, they found the homes and fort deserted. Knowing that a powerful British force bent on exacting revenge was on its way, insurgents and townspeople alike had opted to make themselves scarce. The troops entered Charikar unopposed and set about first looting and then torching the entire settlement. When five days later McCaskill was back in Kabul, one of his officers was heard to remark that the destruction of Istalif and in particular that of Charikar had left such a mark 'as will be remembered for ages'. True enough, for whatever life was breathed back into Charikar after the troops departed for home was violently snuffed out by the dreaded Russians, who finally moved in on Afghanistan 33 years after Britain's withdrawal from India. Charikar sits at the gateway to the Panjshir Valley, the scene of some of the heaviest fighting of the 1979 Soviet invasion.

Extracting the Army of Retribution from Afghanistan was in no way a cut-and-dried affair. The Afghans took pains to ensure that the British troops were given a proper send-off, in the spirit in which they had been received in the country four years before. There was not much left in the way of organized resistance amongst the insurgents, with Akbar on the run and his confederates either hanging from gibbets in Kabul or scattered to the winds. But the tribesmen took every opportunity to harass the retiring columns with *jezail* shot and rocks hurled from their cliff-top sanctuaries. The Ghazis came swooping down in the treacherous passes in a frenzy of killing and plunder, and woe betides the stragglers, who are the Afghans' victims of choice.

Pollock led the advance guard, taking precautions to place pickets at the top of every commanding height until the rearguard had past. But once in the Khyber Pass, Sale neglected to take similar steps to look after his column's safety, and true to form he left things to chance. A party of Afridi tribesmen saw their opportunity and swept down on the rearguard, while another rushed into the baggage column, carrying off scores of mules and camels and inflicting some casualties in the ranks. McCaskill coming up behind suffered a similar costly attack in the pass. Nott handled his troops like a true soldier. He managed to beat the enemy off but not before he had taken 84 casualties by the time his force was back on British India's soil. The general was as usual in a foul mood, and justifiably so, for he was forced to march at a snail's pace to protect an enormous piece of booty, the fabled sandalwood Gates of Somnath, which Ellenborough had instructed him to dismantle at Ghazni and transport back to India. The governor general had issued a document known as the Simla Proclamation,[33] in the same office and four years to the day after Auckland's Simla Manifesto, in which Ellenborough exclaimed to India's Hindu millions that 'the insult of eight hundred years is at last avenged'. The gates were to be returned to their rightful place at the Hindu temple from which they had been looted in the eleventh century by Mahmud, India's first Muslim invader. Nott's political officer, the noted archaeologist Rawlinson, delivered a severe blow to Ellenborough's vanity after he gave the gates a careful examination and pronounced them a fake. The colossal wooden structure was quietly left to rot at Agra. Pollock's parting shot on his march from Afghanistan was to order the destruction of Jalalabad's defences, as well as the fortress of Ali Masjid in the Khyber Pass, the object being to render them useless to the enemy. But just how pointless an exercise in retribution this was is something the British were to learn with great pain less than 40 years later.

The First Afghan War was ended, but it was not yet over. For the moment, no one was passing judgement on the war, which had been a terrible mistake as well as an injustice to Dost

Mohammed and his people. Auckland and Macnaghten had failed to establish Shuja on the throne, and the army commanders who were dispatched to Kabul to prop up this despised puppet had brought upon the Raj its greatest ever military disaster. But on 13 December 1842 such spoilsport observations were far removed from the mind of Ellenborough, who sat astride his charger by the banks of the Sutlej. The governor general was straining to catch a glimpse of his victorious returning army. As the advance guard was sighted approaching the boat bridge that had been laid across the river, with Sale ceremoniously at its head, the 16th Lancers' band struck up a stirring rendition of 'Hail, the Conquering Hero'. This was the signal for a glorious military review and celebrations to commence. The three great generals of the hour, Sale, Pollock and Nott, rode through a great ceremonial arch, behind which stood more than 250 caparisoned elephants arrayed in a two-mile line. It is said that Ellenborough himself took great delight in helping to decorate the beasts. It took ten days for the entire Army of Retribution to return to British territory and on 23 December the festivities commenced, with feasting, military displays and a review of 40,000 troops, in the presence of such dignitaries as the rulers of the Sikh nation and General Sir Charles Napier, who the following year was to add to Britain's Indian dominions with the conquest of Sind. Not since Waterloo, it was said, had British arms cause to celebrate so remarkable a triumph, and it was truly a grand day for Ellenborough and the other civil and military authorities in attendance. There was, however, only one incident to blemish all this magnificence: the elephants, sensing in their wisdom the absurdity of the occasion, refused to trumpet on command.

CHAPTER 5

The Pure Instinct of Dominion

Nearly four decades were to pass before a British army once again marched along the path of conquest into Afghanistan. In 1842, the prospect of a second war with that country would have been received by the British Government as an unthinkable horror. Yet the policies pursued by that same Government made war inevitable, and the seeds of the conflict were sown only a few years after Britain's first disaster beyond its Indian frontier. Nott's former political officer in Kandahar, the noted Orientalist Henry Rawlinson, reflected with acumen shortly after hostilities broke out in 1878 that:

> The announcement of the imminency of another Afghan war has taken the British public by surprise, but to those who have been behind the scenes, and have watched with anxious interest the progress of events on and beyond the Indian frontier, it has been evident for some years that such an issue was almost inevitable.[1]

But as we shall see, Rawlinson's analysis of what he calls the 'two distinct elements of mischief' (the 'intractable character' of the Emir Sher Ali Khan and Russian expansionism) unfortunately do not stand up to scrutiny.

Akbar Khan,[2] who had been appointed vizier by his father Dost Mohammed, died in 1845 at the age of 29, amid quiet speculation that the emir may have had a hand in his ambitious son's sudden demise. In Afghanistan's ruling circles, poisoning, as well as forms of mutilation – in particular blinding[3] – were routine procedures for disposing of rival family members. The backdrop for such deeds was generally a dynastic power struggle of Shakespearian magnitude, as was the case with the founder of the Afghan Empire Ahmad Shah's son Timur, who died in 1793, leaving more than 30 heirs to contest the throne. The fifth eldest, Zaman Shah, imprisoned his brothers, blinding one of them, and was then himself ousted and blinded by his brother Mahmud. Zaman Shah spent the last 40 years of his life chained up in the Bala Hissar's dungeon. Mahmud was in turn blinded and killed by his powerful vizier Fateh Khan, the father of Dost Mohammed. And so it went, right down to our own time, which has seen half a dozen Afghan rulers assassinated in the 50-year period starting in 1929, up to the Soviet invasion of 1979, though strictly speaking these murders were perpetrated more in the spirit of political than fratricidal violence. In 1843, with the restoration of the Barakzai dynasty, the venerable Dost Mohammed found himself in a situation not unlike that of the current Afghan Government nearly 170 years later. He was the undisputed Amir of Kabul, and he held sway over Ghazni, or what was left of the place in the wake of Nott's withdrawal, as well as the equally demolished city of Jalalabad. But his writ did not run beyond those three strongholds. It was ten years before he was able to conquer Afghan Turkistan and occupy Balkh, to which he appointed his son Afzal Khan as governor. At about the same time he recovered Kandahar on the death of his brother, the sirdar Kuhn Dil, and in 1845 Herat was for the first time added to his kingdom.

Dost Mohammed's second reign, which covered a span of 20 years from his return from exile in Ludhiana until his death in 1863, coincided with a phase of unprecedented expansion as well as turmoil for British India. The period between the First and Second Afghan Wars saw British armies in India involved in almost uninterrupted military action, ending with the final dominion of the subcontinent up to the Afghan border. The first territory to fall under British sovereignty during this period was

Sind, a vast, mostly desert land situated between British India and the wild tribal territory of Baluchistan, lying on the vital trade routes to Karachi and up the Indus into the heart of Central Asia. Sind had become a vassal state of the Durrani Empire in 1747, and while it had since shaken off Afghan sovereignty, the local emirs, who numbered 18, kept more than passing ties with their former masters, which posed a very worrying state of affairs for Calcutta.

The trouble began in early 1843, a few weeks after General Sir Charles Napier had been partaking of Ellenborough's Christmas festivities with the returning Army of Retribution on the banks of the Sutlej. Wellington had pulled Napier out of a plum posting as Military Resident of the Ionian island of Cephalonia and sent him to India as the new supreme authority in Sind. The historian Sir Penderel Moon considers this appointment 'an act of folly', unless, he adds, 'it be assumed that to provoke a conflict with them [the rulers of Sind] was positively desired'.[4] Moon answers his own assumption when he acknowledges that Napier arrived in India 'still thirsting for an opportunity to command an army in battle and well fixed in his prejudices'.[5] Napier was sixty and afflicted with rheumatism when he took command of the Army of Sind. Nonetheless, he retained all his strength of character and above all a set of messianic views with regard to Britain's civilizing mission in India. Like many of Napier's contemporaries in position of high office, he was a man possessed of what the Duke of Argyll defined as 'the pure instinct of dominion'.[6] Napier was convinced that Sind, like the rest of the subcontinent, should be annexed to the British Empire, and the sooner the better. Given the local rulers' understandable antagonism to this idea, there was only one way for Napier to achieve his objective, the way the veteran of the Peninsular and Anglo-American Wars knew best. In short, Sind was a volatile and hostile region which, to Napier's mind, needed to be subdued without delay. This eccentric swashbuckler possessed as few scruples as the politicians who had pushed Britain into Afghanistan, but had at least the honesty to avow it.

'We have no right to seize Sind', he wrote in his diary, 'yet we shall do so and a very advantageous, useful, humane piece of rascality it will be.'[7]

The general's first step towards provoking hostilities was to demand the emirs accept a new treaty[8] that would significantly limit their powers, as well as cede more territory to British India. Napier also proposed with sublime Victorian arrogance that the emirs should mint new coinage displaying Queen Victoria's head, as a sign of vassalage to British India. Quite predictably, Napier's ultimatum to the Sind chiefs incited a wave of violent protests, with the city of Hyderabad, one of the largest in Sind, paralysed by an outbreak of anti-British protests. An attack on the British residency at Hyderabad was the general's cue to march an army of 2,200 sepoys to meet a force of 20,000 of the emirs' men at Miani, on the Fuleli River, on 17 February 1843. The Sindis' losses that day are put at 6,000, compared with fewer than 300 for the British side. However suspect Napier's motives for going to war, there is no denying his brilliant leadership in that campaign, as well as the follow-up action at Dabu, eight miles from Hyderabad. In the second battle, Napier's 5,000 sepoys literally decimated the emirs' ragtag army of 25,000 men, the general's force once more taking only 30 casualties. By 14 June 1843 all organized resistance to British power had ceased and in August of that year Napier announced the annexation of Sind to Britain's Indian Empire, allegedly in the immortal one-word telegram: *Peccavi* ('I have sinned').[9]

Napier sailed to England in 1847, having bequeathed 54,400 square miles of territory and more than 2 million new subjects to the East India Company's dominions. His victory in Sind also placed the British in possession of the only remaining strip of Indian seacoast which, until then, had been outside their jurisdiction. The Government now turned its covetous eyes northwards to the Punjab, where since the death of Ranjit Singh in 1839, the annexation of the Sikh kingdom began to occupy the thoughts of the highest colonial and Home Government officials. The Sikhs had remained on ambiguous friendly terms with the British throughout the First Afghan War, resisting the temptation to join in the attack on a foundering imperial power. That was a stroke of luck for British India, which would have had to request reinforcements from the tight-fisted Peel Government to fight a two-pronged war against the Afghans and the Sikhs. The disappearance of Ranjit Singh's commanding presence was not good news, however, because this opened the floodgates to a wave of horrific family assassinations, to the point that by 1844 the collapse of the Sikh political and social order came to be viewed as imminent. The Government's immediate concern was that prolonged anarchy in the Punjab would by sheer force of momentum spill over into British territory.

The East India Company's Court of Directors was alarmed at Ellenborough's Napoleonic proclivities and in 1844, while Napier was still campaigning, the Directors issued an order for the governor general's recall, with the Cabinet's blessing. He began his homeward journey in August, 'leaving behind him a reputation for vainglorious arrogance'.[10] Ellenborough of course saw things through a different prism: his term of office had been crowned first by the victory of the Army of Retribution, and latterly by Napier's victory in Sind which, apart from filling in a gap on the map of British territory, also secured the opening of the Indus for trade. And even greater glories lay in store before Ellenborough's departure for England. Napier followed up his Sind campaign with an attack on Gwalior, a kingdom south of Delhi ruled by the Rajput dynasty[11]. Ellenborough, who all his life had longed to be a soldier, once came close to seeing his dreams fulfilled when he witnessed first-hand the Battle of Maharajahpur, at which the Maratha Army was defeated and Gwalior taken. This secured the rear and the communications of the British Army in the event of war with the Sikhs.

Just as Ellenborough was recalled, the situation in the Punjab had reached boiling point. The succession to the Governorship General of India remained in the family, though passing to a man with the military background Ellenborough had always craved. The person chosen to replace him was Ellenborough's brother-in-law, Viscount Henry Hardinge, a soldier who had campaigned in Spain and Portugal, and distinguished himself at the crucial Battles of Corunna and Vitoria. Hardinge had his left hand shot off fighting Napoleon's troops at Ligny in 1815, but being an exceptionally brave commander, he earned great fame in the ranks by remaining at his post throughout the action. Hardinge arrived in Calcutta in July 1844, knowing full well that war with the Sikhs was on the cards. Intelligence reports from Lahore spoke of widespread disorder and the lack of anyone in the ruling family strong enough to control the ambitious and ruthless chiefs who were struggling to gain supremacy. Ranjit Singh's only legitimate son, Kharak Singh, whom Auckland's sister Emily had correctly termed an 'idiot', had been murdered. Sher Singh, who was next in line for the throne, met the same fate three years later. Tumults and assassinations followed in quick succession. The child heir Duleep Singh ruled with his mother Rani Jindan as regent.[12] The real power rested with the Sikh Army, or Khalsa, but this too was rife with factionalism amongst the military leaders. Disbanding the army to stop the political anarchy was not an option: diverting its energies into war was the obvious answer. 'Like many unsteady regimes', relates one historical account, 'the Sikh government had the choice of war or internal turmoil, and it chose war.'[13]

Six months after Hardinge took up his duties, Commander-in-Chief Sir Hugh Gough, who had the singular habit of wearing a white coat into battle as a rallying point for his soldiers, warned the governor general that it was time to reinforce the garrisons contiguous with the Sikh kingdom. 'He [Gough] pointed out that the British had no field artillery fit to cope with the heavy metal of the Sikhs.'[14] Gough, it may be noted, was eager to see these defences strengthened, for he planned to personally lead his army into any confrontation with the Sikhs. In the end, he was in command at the seven pitched battles fought in the two Sikh Wars. The Khalsa, moreover, was no band of tribal ruffians sniping behind rocks, like the *jezail*-wielding Afghans. Ranjit Singh had ensured his army received professional training along European lines. 'It was a formidable machine', according to British Army historian Philip Mason. 'Ranjit Singh had gone much further than any other Indian leader in adopting European methods. He had borrowed both from the French and the British.'[15] The Khalsa also counted amongst its ranks a number of European soldiers of fortune employed as advisers on matters of military strategy and weaponry. The most eccentric and certainly most outrageous of this group was Colonel Alexander Gardner, an American-born son of a Scottish physician. Gardner was an adventurer who spent years wandering across Central Asia, eventually finding his way to Lahore where, thanks to his ability to read an English instruction manual, he was able to show his Sikh friends how to manage two captured guns and a number of time-fuse shells. Out of gratitude, Ranjit Singh bestowed on him the rank of colonel on the spot. It was all very dashing stuff, however the official Punjab Records give a version somewhat at variance with Gardner's own tales. It would appear that Gardner was a deserter from a European regiment, a bloodthirsty individual who once cut off the nose, ears and right thumb of a prisoner, and who ended his days running a small business in Kashmir, attired in tartan uniform, where he died at the age of ninety-eight – or seventy-six, as indicated by the scratched out entry on his death certificate.

However it was not the ranting of a barmy American in Ranjit Singh's service that gave concern to the Government of India. There was the very real threat of a Sikh army crossing the Sutlej and overwhelming the poorly-defended British positions, an army, it must be emphasized, that had been brought up to world-class standards by genuine European officers like Jean Baptiste Ventura and Jean François Allard,[16] and even the formidable Avitabile, whom we have already met. The anticipated crossing took place on 11 December 1845, and on that same day Gough ordered his cavalry to advance from Ambala. The general himself followed with his infantry the next day. The British were in an uncomfortable position, given that the 30,000 men under Gough's command were widely dispersed along the Punjab frontier. The politicians in Calcutta and Whitehall regarded this as an opportunity – provided the military operations went well – to vastly increase the size of the Empire and in fact fill in the last 'pink bits' on the map of British India. The field commanders pondered the disquieting realities of their situation.

Ferozepore was held by 7,000 men of the 27th Native Infantry under General Sir John Littler, a lifetime Bengal Army man who had commanded a division at the defeat of the Gwalior Army in 1843, where he was slightly wounded and had two horses killed under him. This garrison was the first to come under threat of attack by the Sikh Khalsa, which were led by Lal Singh. Gough gave orders for the different detachments to converge as quickly as possible towards the isolated post, which lay scant miles south of the Sikhs' crossing point. The British brought up a force numbering no more than 12,000 men, who moreover had travelled more than 100 miles in five exhausting days to arrive at the village of Mudki, 20 miles south of Ferozepore. This was too good an opportunity for a staunch old soldier like Hardinge to miss, so the governor general himself joined the advancing troops. While waiting for the many thousands of stragglers to catch up with the advance column, Hardinge and his commanders had their lunch, and presumably their digestion was interrupted by the news that the Sikhs were seen to be approaching. Major George Broadfoot, who had served as Sale's garrison engineer at Jalalabad, rode out to confirm the report. He pointed to a great cloud of dust with the words, 'There, Your Excellency, is the Sikh Army.'[17]

Despite facing a determined enemy, the more professional British troops were able to rout the Sikhs in a pitched battle that lasted until midnight at this first major engagement, in which the 3rd Light Dragoons made a celebrated charge that earned them the nickname 'The Mudki Wallahs'. It must be added that Lal Singh, who headed the Sikh attack, deserted his army and fled the field while the Sikhs stood firm in their order, fighting in a resolute and determined manner. The British infantry gave a splendid account of themselves, taking 17 pieces of the Sikhs' heavy artillery in furious hand-to-hand combat. The

Plate 1 Ahmad Shah Durrani, Founder of the Afghan Empire in 1747.

Plate 2 Dost Mohammed with his youngest son. One of Afghanistan's wisest emirs, Dost Mohammed ruled from 1826 to 1863. He

Plate 3 Shah Shuja.

Plate 4 Lady Florentia Sale, whose diaries revealed the tragic events that led to the Army's destruction on the retreat from Kabul.

Plate 5 Alexander Burnes, the Government's envoy to the Court of Dost Mohammed, was the first high-profile victim of the Kabul uprising.

Plate 6 Akbar Khan was Dost Mohammed's favourite son, and was responsible for the Army's massacre.

Plate 7 William Macnaghten, the Government's chief political representative in Kabul, was murdered by Akbar Khan.

Plate 8 The Last Stand of the 44th Regiment at Gundamuck, 1842. Painted by William Barnes Wollen. Image courtesy of the Essex Regiment Museum, Chelmsford.

Plate 10 Sher Ali Khan, known as the Iron Emir, was Afghanistan's longest ruling king, who occupied the throne from 1863 to 1879. He died in exile after the British launched their second invasion of Afghanistan.

Plate 11 Field Marshal Lord Roberts of Kandahar, the legendary military figure who led the famous march from Kabul to Kandahar, where he defeated the Afghans and won the Second Afghan War.

Plate 12 The Battle of Maiwand turned into a disastrous rout that cost the British more than one thousand casualties. It was one of the few instances in the 19th century of Asian forces achieving victory over a European Army. Image courtesy of the National Army Museum.

Plate 13 A Russian stands in frustration at the door to British India, padlocked out by the Treaty of Gandamak.

Plate 14 After the massacre of the Cavagnari Mission in Kabul, the Afghan is fed a portion of British rule.

Plate 15 Badges in the Khyber Pass of British and Indian army regiments that fought in the Afghan wars and on the North-West Frontier.

Plate 16 A caravan negotiating a narrow pass in Afghanistan. These caravans were also used to transport smuggled weapons to Pashtun fighters.

Plate 17 King Amanullah invaded British India in 1919, touching off the Third Afghan War. The Emir lost the war but claimed victory for having recovered Afghanistan's full independence. He was forced to flee his country in 1929.

Plate 18 The Khyber Pass, the major trade and invasion route between Central Asia and India.

Plate 19 A street in the Peshawar bazaar around the time of the Third Afghan War.

Plate 20 Royal Air Force BE2C biplanes, used to bomb Afghan positions in the Third Afghan War and later against rebellious Pashtun tribesmen on the North-West Frontier.

Plate 21 A newspaper illustration of the RAF bombing Kabul's Bala Hissar fortress in the Third Afghan War.

Plate 22 The border crossing between the North-West Frontier and Afghanistan. Travellers are warned to leave here on the return journey and be back in Jamrud no later than 5:00pm. Image courtesy of the Field Family Collection.

day, however, was not without its tragic events. In this battle, two of the army's most distinguished veterans of the First Afghan War, Generals Sir Robert Sale[18] and Sir John McCaskill, fell alongside more than 800 of their comrades. The Battle of Mudki marked an example of heroism and tenacity across all ranks, right up to the highest levels. Gough himself did not leave the field until two in the morning, while Hardinge did not hesitate a moment to take charge of the 3rd Division when McCaskill fell mortally wounded. There was little rest for the troops, however, for Gough marched in pitch darkness to a position within two miles of the Sikh entrenchments.

By 11 a.m. all was ready for a second action, this time at Ferozeshah. The British had mustered a considerable force of 18,000 men and nearly 70 guns for this action. The Sikhs put up a stubborn resistance to Gough's preferred tactic of hitting the enemy with a massive frontal charge. The guns on both sides cut great swathes in each army's ranks throughout the day. By nightfall, the British found themselves in a critical position. Ferozeshah was rapidly turning into the most savagely contested action fought by the British in India. The battle did not come to an end until 4 p.m. on the afternoon of the 22nd. By a stroke of luck, the Sikh commander Tej Singh missed his chance to win an easy victory by opening up with his guns on the exhausted British troops, who were too weary to mount a bayonet attack and had barely a cartridge left in their pouches.

Hardinge held out little hope for a British success. Thinking the worst was in store for the army he handed his sword, a gift from the Duke of Wellington, and his Star of Bath, to his son, sending him off to the comparative safety of Ferozepore. At this point the gallant old Gough, in his conspicuous white coat, rode out with a single aide-de-camp to draw the enemy's fire. The effect was electrifying: the 3rd Light Dragoons spurred their horses into a gallop and crashed into the stationary mass of Sikh cavalry waiting 150 yards from the British position. Tej Singh judged the situation to be hopeless and withdrew his troops from the field. It can safely be said that this was an exclusively British victory, for the sepoy troops showed themselves reluctant to engage the Sikhs, a foreboding of dark events that were to shake the Empire in a decade's time. As a result of the burden falling on the Queen's regiments, of Gough's 2,415 casualties, fully half were British troops, who normally made up a fraction of the total in engagements involving European and native regiments.

The last battle of what would come to be known as the First Sikh War took place on 10 February 1846. The Sikhs were determined to throw everything they had at the British to stop their advance on Lahore, which was now within the grasp of Gough's force. The Khalsa formed a defensive horseshoe position at the village of Sabraon on the Sutlej. Lal Singh, sensing that the Sikhs were about to take a beating, had his eye cast on future rewards and was secretly supplying intelligence to British officers. He advised Gough and Hardinge to launch a frontal attack on Sabraon and having done so, they put the Sikh artillery out of action in two hours, following up with a three-pronged cavalry and infantry assault on the enemy entrenchments. Tej Singh, who was in command, fled the field, giving the day to Gough's force, but at a cost of 2,400 British and sepoy dead. Two days later the British crossed the Sutlej and established camp at Kasur, a short march south of Lahore.

The question now arose of how to dispose of the spoils, namely the Sikh kingdom that lay prostrate before British arms. One solution advanced by the hardliners, the proponents of what was called the Forward Policy, advocated outright annexation. Hardinge opposed this idea, on the grounds that this would lose British India a useful buffer against potential aggression from Afghanistan, or indeed Russia if the tsar's armies continued their southern march towards the Afghan border. The formula that was eventually adopted provided for an emasculated Sikh state, somewhat along the lines of post-First World War Germany. A treaty was drawn up in March 1846 imposing strict limitations on the size of a Sikh army, the surrender of territory, including Kashmir, to British India, and payment of £500,000 in war reparations. Maharani Jind Kaur continued for the time being to act as Regent of Duleep Singh, while Lal Singh was duly compensated for his collaboration with the title of vizier.

Real power had passed to the hands of the British Resident Sir Henry Lawrence, the soldier who had led the Sikh contingent into Afghanistan with Pollock's avenging army. Lawrence, like Hardinge, was opposed to annexation of the Punjab. His argument was that the Sikhs, whom he knew better than most British officials, would make a potential source of support for British India if they were allowed to retain their homeland and dignity. Meanwhile, Lal Singh and his henchmen were embroiled in a hotbed of intrigues. Lawrence got fed up with conspiracy plots being hatched behind his back and summarily deposed Lal Singh and ordered the young Duleep Singh to be separated from the scheming Maharani Jidan. Then, in 1847, Lawrence was obliged to return home on sick leave, sailing back to England in the company of Hardinge.

The departure of this powerful figure hastened the outbreak of a national revolt. Hardinge's successor James Ramsay, the Marquess of Dalhousie, had been only three months in office when the rebellion broke out in April 1848. Dalhousie and his predecessor shared a fighting spirit, and while the newly arrived governor general lacked Hardinge's military experience, he threw down his gauntlet with a firmness of purpose, declaring 'The Sikh nation has called for war, and in my words, Sirs, they shall have it with a vengeance.'[19] The Sikhs were playing a losing game: only a remnant of their once mighty Khalsa remained to pitch against British troops. By deliberately provoking a conflict, first by sending their army across the Sutlej and now in armed rebellion, the Sikhs fatally assumed that British India's armies were exhausted after four years of warfare in Afghanistan. It was the same mistake the Sikhs' erstwhile friends the Afghans were to make 75 years later, when they invaded India in the belief that the First World War had left the British drained of the will to fight.

On learning of the rebellion, Lawrence hurried back to India, landing at Bombay in December 1848. He went at once to the Punjab to join the army in the field. There he witnessed the bloody Battle of Chillianwala on 13 January 1849, which left the opposing forces in an inconclusive position until little more than a month later, Gough scored a decisive victory at Gujarat. The British infantry advance turned the retreat of the Sikhs into a general rout, the troops throwing away their arms as they dispersed in every direction. Gough's force continued the chase for 15 miles beyond Gujarat and did not return to camp until 10 p.m. that night. The Khalsa was shattered beyond hope of recovery and the once formidable Sikhs tendered their unconditional surrender in March. Dalhousie's impetuousness was that of a 37-year-old placed in charge of an Empire, and this threw him into conflict with Lawrence, who had resumed his duties in Lahore. In the face of the Resident's strong opposition, and the lack of clear guidance from London, Dalhousie ordered the annexation of the Punjab. The shape of British India was now virtually complete, leaving the British masters of India from the Indus to the Bay of Bengal.

But what of this *terra incognita* beyond the Indus? Britain had acquired some knowledge of Afghanistan through Mountstuart Elphinstone's magisterial work on the country, its people and their history, and of course through four years of warfare, though a more attentive reading of Elphinstone might have helped to avoid some of the tragic blunders of the First Afghan War. The annexation of the Punjab left Britain master of a vast tract of territory inhabited, in essence, by Indians, a people the British had come to understand in the course of more than two centuries of conquest and expansion. As for the trans-Indus Vale of Peshawar, the transfer of sovereignty represented a fairly effortless task. Peshawar and the route through the Khyber Pass on which it stood was well-trodden terrain for the British, who had used the city as a staging post for the army in the recent war. With the exception of the historic route into Afghanistan, what lay to the west, in those dark, forbidding mountains bordering Afghanistan, remained an enigma.

Britain was now in close proximity with Afghan tribes, only now these were to be found on the eastern divide of an ill-defined border which placed them nominally in British territory. The border Pashtuns live sandwiched between their Afghan cousins the Durranis and Ghilzais, and the settled, 'civilized' tribes of the Peshawar Valley. These barren hills are home to some 50 tribes, a land so inhospitable that preying on one's neighbours became a way of life and the only viable mode of survival in this desolate environment. This predatory existence can quickly transmute into united resistance to foreign invaders. In 1849, the British were given the same belligerent reception that had been accorded every intruder since the days of Alexander the Great. The British experience on the North-West Frontier could be summarized in a Pashtun proverb that came into use in the nineteenth century: 'First comes one Englishman for *shikar* [shooting], then come two Englishmen to draw a map, and then comes an army to take your land. So it is best to kill the first Englishman.'

This rugged strip of tribal land extends from Chitral, bordering the Hindu Kush to the north, 400 miles south through a tangle of inhospitable hills and mountains to South Waziristan, the homeland of the fearsome Mahsud tribe. At its widest point, it runs from the Kurram Valley on the Afghan border some 250 miles eastward to the city of Abbottabad. The British took direct administrative control of the region between the Indus and the tribal foothills to the west. The Pashtun hills enclave, which is today known in Pakistan as the Federally Administered Tribal Areas, or FATA, was to be left to govern its own affairs under the tenets of Pashtunwali, the tribal code of honour, as it had been for centuries past. This was conditional on the tribes keeping peaceful relations with the British of the so-called Settled Districts to the east. In essence, this meant no raiding, a proviso that unfortunately fell on deaf ears.

Hostilities with the tribes broke out a scant six months from the first army detachments deploying to the Frontier, and the fighting continued unabated for the next hundred years, until Britain's withdrawal from India in 1947. In the years leading up to the Second Afghan War in 1878, the army mounted 40 punitive expeditions into tribal territory, some at brigade level. The Pashtuns' anger at encroachment by the infidels was inflamed by fanatical mullahs who were encouraged and funded by the Afghans, whose leaders never shirked from an opportunity to make things uncomfortable for the British. The Government's Frontier policy swung back and forth over the years, from aggressive expansion to non-intervention, in line with whoever happened to be in power in Whitehall. The Frontier tribes watched this threat to their hitherto independent valleys with growing anxiety.

Far from reaching an accommodation with the Pashtuns in the years of British occupation, the conflicts grew in frequency and intensity until in 1897, the Frontier burst into flames with the unprecedented event of a general uprising by almost all of the tribes. The driving force behind the revolt is obscure, but captured documents clearly show that Kabul had dispatched religious demagogues to the Frontier to preach jihad, thus stirring up the tribesmen's festering grievances over such issues as an increase in the salt tax and the construction of military fortifications in their territory. Some 75,000 troops were at various stages mobilized to subdue the rebellious tribes, in a campaign that lasted two years and which posed a real threat to British sovereignty on the Frontier. The young journalist Winston Churchill, who was covering the war for the *Daily Telegraph*, was hardly overstating the threat when he wrote, 'The whole British Raj seemed passing away in a single cataclysm.'[20]

Elsewhere in India, the period between the first two Afghan Wars was marked by a calamitous event that shook the British Empire to its foundations. The 1857 Sepoy Mutiny falls outside the remit of the present work, except for the role of Afghanistan, or more precisely the absence of any Afghan involvement in that appalling tragedy. The rebellion by sepoys of the East India Company's army broke out on 10 May 1857 in the town of Meerut and quickly spread to other parts of mainly northern India. Other regions under direct Company control, such as the Bombay and Madras presidencies, were largely unaffected by the uprising. Where it did take root, the Mutiny manifested itself with the most wanton violence against Europeans, women and children alike, and equally brutal reprisals by the British.

The causes of the Mutiny were of many origins and complex in nature. Lord Dalhousie had recently annexed several native states, a move that caused resentment amongst their ruling dynasties. The telegraph and railways were taking hold in India, and this represented to the Indians another symbol of British power. The army and civil service still refused to admit Indians of whatever caste into the higher ranks. The Crimean War, and the memory of the Afghan disaster, left the impression that the English were not invincible, and of course, there is the colourful story of the new cartridges allegedly being greased with cows' or pigs' fat, sacred or unclean animals respectively to Hindus and Muslims.

The Government could ill afford to free up hard-pressed British and loyal native troops outside the immediate areas of hostilities. The North-West Frontier had remained calm, though one signal from Kabul could have ignited an uprising by the Pashtun tribes who were spoiling for a chance to rid themselves of the hated Feringhees. That signal never came, for Dost Mohammed proved himself as good as his word. In 1855 he had signed a treaty with the British by which the emir was bound to be 'friends of our friends and enemies of our enemies'. Then in January 1857, the emir travelled in person to Jamrud, at the mouth of the Khyber Pass, to ratify that treaty of friendship, 'declaring as he did that he had made a lasting alliance with the British' and promising to 'keep it faithfully till death'.[21]

The key to victory over the mutineers was Delhi, the cornerstone of British rule in India. The city had fallen into rebel hands, so with a view to releasing British and loyal sepoy troops from the Punjab to join the siege of Delhi, John Lawrence, who had replaced his brother Henry as Chief Commissioner of the Punjab, proposed handing Peshawar over to Dost Mohammed. The plan was to ask the emir to hold Peshawar – the sacred place of the Afghans which he had begged the British to help him recover before the invasion – 'in trust' until such time as the rebellion was brought under control. To this almost unbelievable and insensitive piece of arrogance the Commissioner in Peshawar, Sir Herbert Edwardes, replied that Peshawar would be held, for to abandon it would be fatal. 'It is the anchor of the Punjab, and if you take it up the whole ship will drift to sea', he wrote in his reply, adding:

> As to a friendly transfer of Peshawar to the Afghans, Dost Mohammed would not be a mortal Afghan – he would be an angel – if he did not assume our day to be gone in India, and follow after us as an enemy. Europeans cannot retreat – Kabul would come again.[22]

Edwardes' judgement was absolutely correct: had Dost Mohammed got wind of any intention to withdraw British forces from Peshawar and the lands west of the Indus, Afghan fighters, in league with their Pashtun brethren of the border hills, would be descending on the city in their thousands.

With the relief of Delhi and the suppression of the Mutiny in 1858, the East India Company was dissolved and the Indian Empire was thereafter directly governed by the Crown in the new British Raj. The Afghans held fast behind their border and Dost Mohammed had resisted the temptation to even the score with the beleaguered British, who had years before deprived him of his throne. Had the wise old emir ever suspected that in 20 years' time another British army would cross the border to attack Afghanistan, he might have acted differently.

CHAPTER 6

Chronic Suspicion and Undignified Alarm

India was now a Crown colony, with the Sovereign's authority vested in the viceroy, the figure of supreme Government authority in Calcutta. Charles Canning bridged the old and new orders as India's last governor general and its first viceroy. Though the titles were jointly held, the former designation was retained for administrative purposes, while the office of viceroy took on a chiefly ceremonial character. The Mutiny had drained the Indian coffers of about £42 million and increased military expenditure by another £10 million. In the post-Mutiny reorganization process, the East India Company was abolished and the India Office in London was created in its place, a symbol of the transfer of the real seat of power to the Home Government. The post of President of the Board of Control was also eliminated and its place taken by a Secretary of State for India who now became, in subordination to the Cabinet, the fountain of authority as well as the director of policy in India. Later, the laying of the Red Sea telegraphic cable in 1870, which enabled quick and direct contact between London and Calcutta, put an end to any independent action by the Government of India.

Canning, who served for another four years after the Mutiny, was confronted with the tremendous burden of addressing British India's pressing social and economic problems. For a starter, the Mutiny had left the Treasury with a £7 million deficit. Then there was the military crisis, inasmuch as all but 8,000 of the 128,000 Indian troops in the Bengal Army had been involved in the uprising, making it necessary to reorganize the Indian Army from the bottom. A programme of social and material improvement that had been initiated by Dalhousie was at a standstill and needed to be restarted. Canning's final years in India were devoted to placing his house in order with these tasks before him, a duty he achieved with a reasonable degree of success, being more of a tenacious than a brilliant cast of mind.

As for India's foreign relations, once the Second Burmese War was out of the way Calcutta's attention was inevitably drawn once more to the inscrutable forces stirring beyond the country's north-western border. Canning's greatest achievement was to reject John Lawrence's advice to pull back from the Indus. Such a move would have been an open invitation to Afghan interference and, in the thinking of the Government's Russophobes, would have also unlocked the gates of India at their most vulnerable point to the dreaded Russian bogeyman. Lawrence's proposal found little support and was not pursued. But his policy of non-interference in Afghan affairs held the field for years afterwards, as a reaction to the policy of the 1830s that had proved so disastrous for Britain.

Relations with Dost Mohammed were few and uneventful in the years following the Mutiny. The treaty that had previously bound the Emir in an offensive-defensive alliance with British India had been put to the test a year before the sepoy rebellion when the Persians finally seized Herat, a trophy they had coveted for decades past. This act of aggression was treated as a *casus belli* by Calcutta and war was declared on Persia, as always with an alarmed eye on Russia skulking somewhere behind the political instability. Herat was only nominally a part of Afghanistan at the time and held to be 'an important element in the defence of British India against the machinations of Russia'.[1]

The British forces were commanded by General Sir James Outram, who after the war was speedily summoned to India where he performed exploits of great brilliancy in the Mutiny at the head of two divisions of the Bengal Army. Outram fought the Persian war in two theatres, on the southern coast of Persia near Bushehr and in southern Mesopotamia, now Iraq. The Herat affair was a miserable little campaign lasting all of six months. It ended in humiliation for the Persian Army and with a pledge from the Qajar rulers in Tehran surrendering their claims on Herat and any other part of Afghanistan. What stood out as significant for Anglo-Afghan relations during this conflict was the reaffirmed treaty, already mentioned, under which in time-tested tradition Dost Mohammed was given £10,000 a month for the duration of the war, in exchange for an undertaking to keep a body of troops in readiness to defend his territory and to allow British officers access to Kabul, Kandahar and Balkh.

In May 1863, at the age of 80, Dost Mohammed set out with his army on an expedition to capture Herat. The objective was to bring Herat formally into Afghanistan under the Emir's dominions, since until then the city-state had been Afghan only in name. Herat fell on 26 May and a fortnight later Dost Mohammed died suddenly in the midst of victory, having played an illustrious role in the history of Central Asia for 40 years. It is a tribute to the emir's integrity that despite the shabby treatment he had suffered at the hands of the British, he remained a loyal ally throughout and kept his word. Fraser-Tytler, who was a long-serving diplomat in Kabul, always maintained that the Afghans required careful and sympathetic handling. He pointed out that the rulers of that country had always carried out their undertakings to the best of their ability. Indeed, Dost Mohammed was the first but by no means the last of the emirs to prove that beneath an enigmatic character, there abides a core of constancy of purpose, which has more than once stood their allies in good stead. Fraser-Tytler even compares the Afghans to the Scottish Highlanders of centuries past, from which he descended, with 'the same love of freedom, the same freebooting instincts, and the same natural courtesy and love of hospitality'.[2]

Dost Mohammed was a prolific progenitor, as well as a powerful ruler of men. He left 16 sons, of whom the third oldest, Sher Ali, had been designated to be his successor, on the ground that he was born of Dost Mohammed's second wife, a Durrani. As might be expected, his two older brothers, Afzal and Azam Khan, though less nobly born, had ideas of their own concerning who should sit on the throne of Kabul. John Lawrence, later raised to the peerage as Lord Lawrence, accepted with some reluctance Prime Minister Lord Palmerston's offer to succeed Lord Elgin, who had died after only a year in office, as viceroy. Elgin was never able to cope with the Indian climate and he collapsed and died of physical exertion on a trip through the Himalayas in November 1863. Lawrence championed the policy of 'Masterly Inactivity' and he made his position clear to the Secretary of State for India Lord Ripon, stating his intention to maintain an attitude of strict neutrality, leaving the Afghans to choose their own rulers. In this vein, the British Government adhered to a policy of vigilant though inactive observation, as Sher Ali and his brothers indulged in the Afghan tradition of settling a dynastic dispute through wholesale fratricide. The fierce contest for power lasted more than five years, throughout which Britain stood aloof on the sidelines, prepared to recognize whoever emerged victorious from the struggle.

This stand-back policy advocated by Lawrence and sanctioned by the Home Government encouraged Sher Ali's two rivals to contest their father's wish for succession. It was an unpredictable tussle, with Lawrence at one point recognizing Afzal, when he succeeded in capturing Kabul, as the rightful emir. 'But Afzal was unable to hold his position. Sher Ali finally triumphed in 1868, driving his rivals into exile. He was promptly recognised as Emir by Lawrence three days before the latter's successor, [Lord] Mayo, landed at Bombay.'[3] Lawrence's logical and well-meaning policy of Masterly Inactivity was a mistake, for it gave Sher Ali no cause for goodwill towards the British. However, had the Government reversed their policy and stepped in to support Sher Ali's cause, it would have made the British *persona non grata* with other important Barakzai factions. None of the contenders seriously expected Lawrence to back their claims with deeds as well as words, and taking sides in Afghan family rivalries is a dangerous game that carries a high risk of supporting the wrong faction. It is almost impossible, in particular at arm's length, to foresee where the knife will fall.

By 1868, those who had thrown in their lot with the usurpers ruling in Kabul were in for a surprise. Sher Ali was back and victorious, and while grumbling about the lack of British support in the rightful heir's hour of need, he was also aware that his best hope lay in keeping cordial relations with the supreme power in Asia. Lawrence sent Sher Ali a letter of congratulations on recovering his kingdom, in which he hoped the emir would be able to 're-establish and consolidate' his authority. Whichever of the Barakzai pretenders emerged victorious was of little consequence to the British. The overriding concern was to be able to rely on a strong and stable regime in Kabul. For Lawrence and the Home Government it was a matter of let the best man win, provided he showed himself sympathetic to British interests in the region. The hawkish proponents of the Forward Policy, who favoured the steady expansion of the Indian Empire, and those who upheld the doctrine of Masterly Inactivity held to one common principle: that Russian influence should be excluded from Afghanistan at any cost. In 1838, Macnaghten's and Auckland's agitation over a Russian thrust through Afghanistan to attack India was a chimera. Thirty years later, the Russophobe school of politics might have appeared to be making more sense.

Russian expansionism in Central Asia began to gather pace from the end of the First Afghan War. Tashkent was stormed and occupied in 1865. Khudjand in present-day Tajikistan, and Djizak and Samarkand in Uzbekistan, fell in quick succession over the next three years. Khiva was taken in 1873 and three years later the fall of the Uzbek city of Khokand completed the annexation of the turbulent khanates. The Forward Policy hawks were convinced that Russia's rapid drive south and east was a prelude to ultimately using Persia and Afghanistan as a staging post for invasion. A less frenzied interpretation held that St Petersburg saw the khanates as a threat to stability on their unsettled borders, which in essence was no different to Britain's fears for security on India's North-West Frontier. Percy Sykes, a soldier-historian with vast experience in the region, maintained that the 'declared object' of Russia was the opening up of a trade route to Central Asia, 'a policy which was partly inspired by the desire to gain contact with the province of Turkestan to the west'.[4]

At the same time, it is useful to bear in mind that Russia was keenly aware of the spread of British hegemony within its Indian Empire, and the defensive treaties that had been signed with Afghan rulers past and present. The Duke of Argyll challenged this one-sided view of events when he referred to British India's conquests after 1842. 'During the forty years which have elapsed since the First Afghan War', he wrote, 'we have conquered and annexed provinces containing many times more millions of men than exist in all the khanates of Central Asia between the Volga and the Wall of China.'[5] These great strides brought into the Empire the princely state of Oudh, with 11 million people, the Punjab, with more than 17 million, and Sind, with a population in excess of 2 million. In less than four decades, British India had absorbed and conquered territories with a population of upwards of 30 million. While these advances were not as gigantic as Russia's, in terms of square miles of territory annexed, they were 'gigantic in the resources they have opened up and in the treasures of which they have put us in possession'.[6]

From the First Afghan War to 1876, the policy of non-interference in Afghanistan's internal affairs was faithfully observed by six governors general and viceroys, from Lord Dalhousie to Lord Northbrook, and five secretaries of state, from Lord Halifax to the Duke of Argyll. After Sher Ali recaptured his throne in 1868, British relations with Afghanistan could be best defined as 'watchful', inasmuch as everyone in Government held a guarded optimism that Britain had found a reluctant ally who might at least be able to maintain a semblance of order in his kingdom. Sher Ali granted Britain permission to maintain a *vakeel*, or native ambassador, at Kabul to represent the Government of India and to act as a listening post on political developments. But trifling a matter though it may seem, the bone of contention that triggered the second war was the emir's resolute refusal to allow European officers to be stationed in any part of Afghanistan. Sher Ali's wishes were not challenged for the moment, on the understanding that this protocol would apply equally to any other European power, specifically with Russia in mind. If the Tsar refrained from intriguing in Afghanistan, Britain's hawks would remain vigilantly poised on their perches, keeping a sharp though neutral eye on the situation.

Russia's conquest of Khokand in 1876 brought the Tsar's armies to a position only 250 miles from British India's northern frontier. By that year, virtually all of Uzbekistan was in Russian hands, right up to the Afghan border. Bokhara and its surrounding territory marched with Afghanistan along a northern frontier that no one had yet accurately demarcated, and in this Calcutta as well as the Home Government found cause for great distress. Britain was no longer prepared to sit back and passively contemplate a Russian ink stain spreading unchecked across the map of Central Asia. But what were St Petersburg's intentions? Was Britain to take the Russian Government at its word, that the only strategic aim of this massive deployment of troops was that of holding in check the wild tribes of the khanates? Nearly a century and a half later, this question continues to provide a lively subject for debate, and the Forward Policy versus Masterly Inactivity remains alive and well in academic venues and gatherings of military and diplomatic historians. There is little evidence to support the hysteria over Russian invasion plans during Sher Ali's reign in Kabul. These fears, however, touched off a war with Afghanistan that cost well in excess of 15,000 casualties, so it is a question worth examining in some detail.

British India in the mid-nineteenth century was in the grip of what one statesman described as 'chronic suspicion' and 'undignified alarm'.[7] A report in *The Times* in early 1873 had helped spread a wave of disquiet in Whitehall. The paper's Berlin correspondent had picked up a disturbing story from the Augsburg *Allgemeine Zeitung* of a memorandum sent to the Tsar by General Alexander Duhamel, the former Russian envoy in Teheran. In this memo Duhamel spoke of Russia's military build-up on its eastern frontiers at the end of the eighteenth century, which he wrote off as 'a harmless measure', as the military preparations that were in progress at the time came to nothing. The general then went on to list a number of possible invasion routes into India that were open to Russia, from Merv, Bokhara and Khokan, along with an analysis of each one's relative advantages. 'It is incumbent upon Russia to consider whether she has the means to touch England in India . . . or at any rate to force her to concentrate an army in Asia, and thereby lame her action in Europe.'[8] It looked like the Forward Policy hawks had been handed a powerful piece of ordnance with which to reinforce their warnings of Russian aggression. The only flaw was that the reporter had lifted a story about a memorandum from a general who had last seen active service in Moldavia 25 years earlier and who had been writing in 1854 when Britain and Russia were at war in the Crimea.

St Petersburg went to considerable lengths to keep Britain guessing as to Russia's Central Asian ambitions, hence the British were determined to settle Afghanistan's frontiers in order to set the boundaries of what would constitute a violation of Afghan sovereign territory. Calcutta pressed the Russians to acknowledge Sher Ali's claims to lands up to the Oxus, and including Badakshan and Wakhan in the country's extreme northern reaches, as well as Afghan Turkestan (Kunduz, Khulm and Balkh) and several internal districts. Russia's Foreign Minister Prince Gortchakoff gave an evasive reply to these demands. The crafty statesman alluded to the two imperial Governments' wish to secure peace and friendly relations through an 'intermediary zone' that would keep their respective possessions from touching at any point. In Russian diplomatic language this implied, *ipso facto*, that the Tsar claimed the right to absorb any territory lying outside Afghanistan's agreed boundaries. There was a dispute over the limit of Badakshan and Wakhan, which the Russians claimed should be regarded as independent.

No sooner had Britain begun to press its position on Afghanistan's northern borders than the Russians did the unexpected by agreeing that 'such a question should not be a cause of difference between the two countries'.⁹ What was shaping up as a bitter diplomatic wrangle suddenly evaporated in a puff of smoke. So, as Gortchakoff had proposed, in 1873 Britain and Russia concluded a treaty confirming Afghanistan as a buffer zone between both empires. This of course left British diplomacy in a state of disarray with regard to Russia's genuine aspirations, which is precisely the game St Petersburg was playing. For barely was the ink dry on the treaty when the Russians annexed Khiva, despite repeated denials of any intention to do so, and then wheeled out the term 'intermediary zone' as a justification. Three years later Russia repudiated this 'intermediary zone' treaty as 'impractical' and sent its troops to occupy Khokand. Now there was nothing but Afghanistan separating the empires of Tsar and Queen in Asia.

February 1874 saw the return to power of the Conservatives under Disraeli, one of Britain's most fervent advocates of the Forward Policy. Disraeli's intolerance of Russian adventurism in Central Asia was legendary. When Constantinople was threatened with Russian occupation during the Russo-Turkish War, Disraeli wrote to Queen Victoria that 'in such a case Russia must be attacked from Asia, that troops should be sent to the Persian Gulf, and that the Empress of India should order her armies to clear Central Asia of the Muscovites, and drive them into the Caspian'.[10]

It was not long before the then viceroy, Lord Northbrook, tendered his resignation in frustration over Disraeli's belligerence towards Afghanistan, and the heavy-handed control imposed on him by Lord Salisbury, the Secretary of State for India. This was Disraeli's chance to designate a more acquiescent candidate to Calcutta and the man selected for the job was not an India hand or even a politician, but a poet whose heart was rooted in the literary salons of London's Mayfair, and whose diplomatic experience was confined to legations in St Petersburg, Vienna and Portugal. Edward Robert Bulwer Lytton, afterwards Lord Lytton, was forty-five and a much admired figure in the Foreign Office when he was offered the viceroyalty of India. Disraeli was certain he could bring Lytton into the Forward Policy fold and in fact in that same letter to the Queen he stated, 'We have a good instrument for this purpose in Lord Lytton, and indeed he was placed there with that view.'[11]

Disraeli offered Lytton the post on 23 November 1875. Lytton replied from Lisbon eight days later, explaining that while no man had ever been 'so greatly or surprisingly honoured', his delicate state of health might render him unsuitable for so demanding a task. His remonstrations were to no avail: Disraeli had chosen his candidate and he was not accepting any excuses on health grounds. He wrote to Lytton on 20 December, 'We have carefully considered your letter, and have not changed our opinion. We regard the matter as settled.'[12] Lytton sailed for Calcutta with his family on 20 March 1876 and no sooner had he set foot on Indian soil, the man who was destined to preside over the Second Afghan War was swiftly converted to the Disraeli line, having pronounced the policy of Masterly Inactivity to be a failure. Lytton's espousal of a hard-line policy on Russia and Afghanistan was set in stone after the new viceroy devoted some careful study to two events that had occurred before his arrival in India.

The turning point for Britain's relations with Afghanistan came in 1873, when Lytton's predecessor Lord Northbrook had invited Sher Ali to Simla to confer on matters concerning British-Afghan relations. The emir declined to go himself and sent an envoy, Noor Mohammed Khan, as his representative, with a consoling message that as ruler of Afghanistan he placed no confidence in Russia and only asked Britain to help him defend his kingdom against foreign aggressors. Sher Ali's decision to deal with the viceroy through an emissary instead of making the journey to Simla in person found its justification in the Government's reply. The Duke of Argyll telegraphed Northbrook with instructions to offer the emir some money, a shipment of rifles and advice, and that was it. The seeds of discord had been sown: Sher Ali did not try to disguise his disappointment at Britain's refusal to negotiate a defensive alliance against Russia. He communicated to Northbrook in deliberately vague terms that European travellers of any nation were not welcome in Afghanistan. The day of Noor Mohammed's departure from Simla signalled the rupture of friendly relations with Kabul.

The second event that transformed Lytton's thinking was a remarkable proposal put forward by Russia's General Konstantin Kaufmann. This influential military hero recommended to Lytton a bilateral direct line of communication between the Russian authorities and the viceroy, to promote good relations with regard to Central Asian affairs. Kaufmann furthermore declared that 'England and Russia . . . had in Central Asia a common interest and a common foe'. This shared interest, he explained, was Christian civilization, while the common foe he defined as Islamism.[13] Kaufmann then quaintly suggested that all the states between India and the Russian possessions in Central Asia be unceremoniously disarmed and divided between the two Western empires. Lytton saw through this flimsy attempt to leave Afghanistan defenceless against Russian aggression and his reaction was firmly in the negative:

> The British Government would tolerate no attempt on the part of General Kaufmann to obtain an influence in Afghanistan . . . and we should absolutely refuse to co-operate with Russia in any anti-Mohammedan crusade, such as that which had been suggested.[14]

Furthermore, he told Kaufmann that Britain regarded Afghanistan as lying within its sphere of influence and would defend her with British power against aggression by any foreign state.

The Forward Policy embraced by Disraeli, Secretary of State for India Lord Salisbury and Lytton was in the ascendant. This hawkish triumvirate was almost single-handedly setting Britain on a collision course with Afghanistan, and as late as August 1878, three months before the outbreak of hostilities, Parliament had no suspicion that war was impending. Extraordinary as it may seem in the run-up to a major international crisis, in the 15-month period between February 1876 and May 1877 not a single dispatch from the Government of India on Afghan affairs appears in the records. It cannot be doubted that there were frequent secret communications between Lytton and Salisbury, but the omission of official reports indicates a system of government wholly outside the law. Less than a fortnight before British troops marched into Afghanistan, a cross-party commission was formed by MPs who were largely opposed to Britain entering into another conflict, particularly one waged on unsubstantiated charges of Russian interference in Afghanistan. 'There was, and is, no evidence whatsoever that Sher Ali thought of inviting Russian aid, or that the Russians were intriguing with him', the commission's voluminous report declared.[15]

Salisbury was striving to persuade the emir to accept British agents in his country, initially in Herat and Kandahar, and finally with the demand of a permanent resident in Kabul. The previous viceroy, Northbrook, had seen no strategic merit in imposing unreasonable demands on Sher Ali. He contended that the bi-weekly intelligence sent by Ata Mohammed Shah, the Government's long-standing and trustworthy native agent in Kabul, for the most part provided an accurate account of political developments at the emir's court. It is ironic indeed that one of the people concurring in this opinion was Sir Louis Cavagnari, who opposed sending a British agent to Kabul. Cavagnari was the British son of an Italian general in Napoleon's Grande Armée and was serving at the time as Deputy Commissioner at Peshawar. He was later sent to Kabul as Lytton's British Resident, with tragic results for Cavagnari and Britain.

Salisbury was less inclined than Lytton to coerce Sher Ali into capitulation with the Kaufmann-inspired threat of an Anglo-Russian military alliance. The secretary of state was alarmed by the proximity of Russia to Afghanistan, geographically as well as in her political dealings with the emir. He envisaged this danger could take one of three forms: Russia might bully Sher Ali into submission, provoke internal disorder or simply invade the country. Whatever course of action Russia might choose – and it was never established that she contemplated any action at all – the Government of India must at all costs station a British Resident in Kabul. Salisbury argued that only the deployment of a European could ensure a flow of reliable intelligence and promote Britain's best interests at court. Sir Henry Rawlinson, a member of Salisbury's council, raised the tone of belligerence a few notches when he advocated the seizure of Kandahar and Herat as a measure to safeguard the integrity of British India's frontier.

Lytton's brief was therefore to negotiate an agreement with Sher Ali to accept a British Resident at his capital. If the emir continued to stand firm, Salisbury stated, he should be 'distinctly reminded that he is isolating himself, at his own peril, from the friendship and protection it is his interest to seek and deserve'.[16] Simply put, Sher Ali could in such circumstances no longer count on British subsidies, or the recognition by Britain of his younger son Abdullah Jan as his chosen successor, or any material support against foreign intrusion in his country. 'These threats', the parliamentary commission concludes, 'would be taken, certainly by a ruler of his [Sher Ali's] class, and probably by any other ruler, to mean war.'[17] Sher Ali's reasons for rejecting Britain's demands were threefold: he could not guarantee the safety of a European in his capital, European officers might make demands that would give rise to conflict, and finally if the British were to send an agent, Russia would surely follow with a similar claim.

Lytton was losing patience with the unreceptive emir. The viceroy found no lack of support from proponents of the Forward Policy, in particular from Sir Bartle Frere, a former governor of Bombay who, as High Commissioner to South Africa, had declared war on the rebellious Zulus. Frere dismissed the claim, put forward on former occasions when this step was urged on the Government of India, that 'a British envoy and his attachés would not be safe from attacks on their lives by fanatical or ill-disposed persons, and that the ruler of the country could not guarantee their safety'. He added:

> I have never believed in the validity of this objection and I should consider it quite chimerical, unless it were formally stated

by the ruler himself. In that case I should point out the absurdity of his calling himself the ruler of a country where he could not ensure the safety of an honoured guest.[18]

In October 1876, Lytton sent a letter to Sher Ali, remarkable in its undiplomatic candour, stating that Britain's only interest in maintaining the independence of Afghanistan 'is to provide for the security of our own frontier'. Lytton continued:

> But the moment we cease to regard Afghanistan as a friendly and firmly allied State, what is there to prevent us from providing for the security of our frontier by an understanding with Russia, which might have the effect of wiping Afghanistan out of the map altogether?[19]

Had it been only a matter of persuading Sher Ali of the benefits of agreeing to a European resident in order to retain Britain's support in the event of attack by a foreign power, the row might have continued to simmer at the diplomatic level. But once the Government discovered that the emir, whilst holding out against pressure to accept a British agent, had, despite assurances to the contrary, allowed a Russian mission at Kabul, his fate was sealed.

Sher Ali could hardly be censured for fearing Britain as a growing menace to his sovereignty, and Lytton did nothing to allay his apprehension. In November 1876 British troops occupied Quetta in force,[20] dredging up memories of 1838, when this had been the first step in the invasion of Afghanistan and the deposition of his father Dost Mohammed. About the same time, army units were moved in the direction of Afghanistan for the purpose of erecting a bridge across the Indus. Confronted with these aggressive military deployments, and threats of 'wiping Afghanistan out of the map', how could Sher Ali not have been expected to view the British Government with anything but deep anxiety?

One need only consider a letter sent from Salisbury to Lytton, which amounted to an unqualified endorsement of the viceroy's hard-line approach to the Afghan crisis. If Sher Ali maintained what the Government termed 'an attitude of isolation and scarcely veiled hostility', Lytton was at liberty to abrogate any existing treaty obligations and adopt whatever measures he judged appropriate for the protection of British India's North-West Frontier, 'without regard to the wishes of the Emir Sher Ali or the interests of his dynasty'.[21] The parliamentary commission described the last sentence as 'a distinct note of war'.[22]

On 7 June 1878, Lytton telegraphed Lord Cranbrook, who had recently taken over as Secretary of State for India from Salisbury, the new foreign secretary, with the bombshell that Sher Ali had accepted a Russian mission at Kabul. The Government's worst nightmare had come true: the Russians were in Afghanistan, though this uninvited mission had come not at the emir's behest, but by sheer force. Sher Ali wanted no part of it, for he knew that a Russian presence in his capital would bring down the wrath of the British Government. In dire alarm, he wrote to General Kaufmann, declining to receive the Russian mission. But it was too late – the Russians would not turn back, and on 22 July General Nikolai Stoletoff rode into Kabul at the head of a large delegation of civil and military officials.

In the meantime, peace had been secured in Europe by the Congress of Berlin, which reorganized the countries of the Balkans in the wake of the Russo-Turkish War. Now that Russia had a free hand to operate in its Central Asian sphere of influence, it would not be advantageous to further antagonize the British by keeping a military presence in Afghanistan, so General Stoletoff was recalled as abruptly as he had arrived. He left Kabul on 24 August, but Lytton was still determined to take action. He wrote to Cranbrook:

> Neither the withdrawal of the Russian Mission, nor any assurances on the part of Russia, will cancel the fact that a Russian Mission has been well received at Kabul, and that Russian officers have had full opportunities of instilling into the minds of the Emir and his Councillors distrust and dislike

towards England, belief in Russia's power and destiny, and hopes of assistance against us from that country.[23]

Lytton was not of a mind to wait for further Russian 'provocations' or, for that matter, excuses from Sher Ali. He advised the emir that regardless of his views on the subject, Britain had deputed an envoy to Kabul. Moreover, the British Government would consider a refusal of free passage and safe conduct of this mission an act of open hostility. The envoy chosen by Lytton to lead the delegation was General Sir Neville Chamberlain, commander-in-chief of the Madras Army, who was instructed to depart Peshawar for the Khyber Pass without delay. This left Sher Ali in a state of great distress – to refuse Chamberlain's mission was to invite war with Britain, while to allow him to proceed would, he believed, give the Russians ground to impose another mission on Kabul. The emir's anguish was compounded by a dynastic crisis, for Lytton's threats coincided with the death of Abdullah Jan, Sher Ali's designated heir.

It must be acknowledged that Chamberlain's was not the most judicious of appointments. The general had taken part in the storming of Ghazni with the Army of the Indus and in 1842 he was with Nott's force in the capture and destruction of Istalif. Chamberlain's military record in Afghanistan was not likely to endear him to the son of Dost Mohammed.[24]

Chamberlain took a party of 200 troops to the Khyber on 21 September 1878, establishing his camp at Jamrud, the village at the entrance to the Khyber Pass that marked the limit of British territory. In view of the hostile reception he anticipated receiving from the Afghans, his strategy was to send ahead a smaller, less intimidating detachment under the command of Major Louis Cavagnari, escorted by a party of Khyber tribesmen who were in the Government's pay. The 'Envoy', as Chamberlain was now presumptuously referred to in official dispatches, established his camp at Jamrud. His reasoning was that if Cavagnari were turned back at Ali Musjid fortress, where the Afghan forces were in waiting, this would entail less of a loss of prestige than having to turn round an entire column.

Cavagnari, taking with him 24 soldiers, was duly escorted to within a mile of the fortress where, as expected, the Afghan General Faiz Mohammed Khan courteously but firmly warned him that if his party advanced a step further, his troops would open fire. Faiz Mohammed sent Cavagnari a message saying he was prepared to meet him at a ruined tower in the stream bed, where he was instructed to hold fast until summoned. Casting an anxious eye at the tribesmen, all of whom were armed to the teeth, lining the ridges on either side of the road, Cavagnari decided to risk it, and he boldly moved up to the tower.

Faiz Mohammed, visibly angered by Cavagnari's insolence, made his appearance surrounded by a throng of fierce-looking, bearded levies. Cavagnari went directly to the point: was the emir prepared to guarantee the safe passage and proper treatment of the Chamberlain mission during its journey to Kabul, or not? Faiz Mohammed's reply was equally plain-spoken: the reply was No. The general explained that without orders from Kabul, he was powerless to allow the British to proceed. Cavagnari saw no point in prolonging the interview. 'I only came to get a straight answer from you, will you oppose the passage of the Mission by force?' he asked. The general replied, 'Yes, I will, and you can take it as a kindness . . . that I do not fire upon you for what you have done already.'[25] Cavagnari rejoined the camp at Jamrud on 22 September and the Chamberlain mission returned to Peshawar. No sooner was Lytton informed of the mission's failure, he fired off an urgent telegram to Cranbrook in London: 'Ordinary diplomatic action is, of course, exhausted, and we must immediately adopt other measures.'[26] Lytton was fully prepared for those other measures. What was contemplated was a military strike to achieve one or other of two results, 'the unconditional submission of the Emir, or his disposition and the disintegration of his kingdom'.[27]

The military plan was to reinforce Quetta, the gateway to Kandahar, with 6,300 troops and 27 guns. Another smaller force was to be deployed to the Kurram Valley in preparation for the invasion, threatening Jalalabad and Kabul. Before hostilities broke out, the number of troops in each column was strengthened and a third was formed to march through the Khyber. On the political front, Cavagnari was actively engaged in talks with the Khyber tribesmen, with the aim of bringing, or more accurately bribing, them into the British camp to ensure the army's unmolested passage through the Khyber Pass. These negotiations achieved far more than the desired results, for in the longer term the deal struck with the Khyber tribes resulted in a major shift in Government strategy for protecting the North-West Frontier, and also brought to centre stage one of the most remarkable characters in Frontier history, Warburton of the Khyber.

On the suggestion of the Home Government, which was always more sensitive than Calcutta to Britain's image abroad, Lytton was directed by Cranbrook to issue an ultimatum to Sher Ali. The emir was sent a self-righteous and tedious recapitulation of the events leading up to the crisis: the acceptance of a Russian mission to Kabul, the refusal to accept a British Envoy, the rebuff of Cavagnari's party in the Khyber Pass, and finally a date, 20 November 1878, by which time Sher Ali would either yield to the British demands, or prepare for war. There was now no turning back from the brink of war, a conflict founded solely on Sher Ali's acceptance of a Russian mission in Kabul and his refusal to admit a British delegation to counter Russia's interference in Afghanistan.

From a commonsense standpoint, it never occurred to Lytton and the Home Government that their real quarrel was with Russia, not Afghanistan. Moreover, the abrogation of Britain's treaty obligations with Kabul left the emir free to accept representatives from any country of his choosing. Logical analysis was of little interest to men who were beyond the reach of ordinary reasoning, and whose thoughts were exclusively focused on military strategy for the forthcoming invasion. 'The cause assigned for the war is not Sher Ali's offences, but our jealousy and fear of the Russians', concludes the parliamentary commission. 'Nothing that Sher Ali could have done, short of a complete surrender of his independence to us, would have satisfied our Government.'[28]

Lord Lytton had effected a remarkably seamless transition from accomplished linguist and poet to eminent wartime statesman, or so he must have fancied himself. On 21 November, the day after the ultimatum to Sher Ali had expired, Lytton began his dispatch to Cranbrook with a thundering *Alea jacta est!*, 'The die is cast!', the phrase uttered by Julius Caesar on crossing the Rubicon in 49 BC.[29] He informed the secretary of state that he had instructed the army to cross the Afghan frontier at three points, the Khyber, Kurram, and Quetta. Thus, in November 1838, the Army of the Indus was assembled at Ferozepore, preparing to march on Afghanistan to depose the Emir Dost Mohammed. In November 1878, the British Army once more stood poised on the Afghan frontier, with the objective of dethroning Dost Mohammed's son Sher Ali. The lessons of history, as the poet T. S. Eliot once said, are timeless.

The Khyber Pass invasion route was placed under the command of Lieutenant General Sir Samuel Browne, the one-armed soldier who famously designed the Sam Browne belt.[30] Colonel Sir Robert Warburton was at that time Assistant Commissioner at Peshawar. Warburton was half-Afghan by birth, a fluent Pashtun speaker and the army's most informed soldier on tribal ways. Browne knew that the predatory tribesmen posed the greatest single danger to leading his 16,200-strong Peshawar Valley Field Force with 48 guns in tow into the treacherous defiles of the Khyber.[31] That was when Warburton came up with an idea: why not turn poachers into gamekeepers, by raising a tribal levy officered by British soldiers to protect the troops on their passage through the pass? So came into being the Khyber Jezailchis, named after the long-barrelled flintlocks carried by the tribesmen who formed the ragtag embryo of what was to become the Khyber Rifles, the first of the Frontier forces that today continue to guard the tribal lands of the Pakistan-Afghan border.

Warburton's father was one of the officers taken hostage by Akbar Khan in 1842, when the army retreated from Kabul. He was fortunate to escape death, for Warburton Snr. had eloped with an Afghan princess, a niece of Dost Mohammed who happened to be already married to an Afghan nobleman at the emir's court. Akbar Khan failed to find and kill the woman who had defiled Islam by running off with this Feringhee, and she managed to stay a step ahead of him by taking shelter in homes of friends until the end of the war. Robert Warburton grew up much like the young hero of Kipling's novel *Kim*, with divided loyalties between his British forebears and his Pashtun friends, until he emerged from the Woolwich Military Academy a staunch imperialist who was later befriended by Queen Victoria and the Prince of Wales. He never lost his admiration for the Pashtuns of his youth, a sentiment that aroused suspicion in some political circles, in which Warburton was acknowledged as the right man for the job, but, as a 'half-caste', not quite clubbable.

When he was appointed Political Officer of the Khyber, Warburton at times had to suffer shabby treatment from the 'politicals', a situation that came to the attention of Lord Salisbury, who was then prime minister. Salisbury wrote to Queen Victoria after Warburton's death from chronic dysentery in 1899, complaining about this unjust hostility towards a loyal servant of the Raj. 'The racial prejudice under which Sir Robert Warburton suffered is deeply regretted, but it is too deeply imbedded in the British official nature to be rooted out.'[32]

The orders were for the three columns to advance at daybreak on 21 November. In late September, a brigade under Major General Michael Biddulph was ordered to reinforce Quetta, an appointment somewhat open to discussion given that he lacked any experience of active service in India. Biddulph was sent as an advanced portion of the main Kandahar Field Force whose commander, Lieutenant General Donald Stewart, Lytton urgently summoned from England by telegraph to take up his post. Biddulph's men were rushed across from Multan, in the parched southern Punjab, over a waterless route that had been reported impassable for wheeled artillery. Many men fell ill by the wayside, suffering intense heat and sandstorms, while hundreds of transport animals were lost on the march. Many of the hundreds of camp followers accompanying the column, being ill-clothed, fell victim to pneumonia. Biddulph's exhausted troops reached Quetta on 9 November.

The immortal general (later Field Marshal Lord Roberts of Kandahar, or 'Little Bobs' to his troops) Frederick Roberts was given command of the Kurram Field Force. This formidable though diminutive soldier recognized with some reluctance the military wisdom of raising a third column to attack through the Khyber, which would relieve the pressure on his Kurram force. But it was not a happy decision for Roberts, who saw Browne as the gainer at his expense. 'This third column was not quite so desirable', he wrote in his memoirs, 'as it involved the withdrawal of three of my most efficient regiments, and the transfer of a large number of my transport animals to the Khyber for its use.'[33] Small as he was in stature, it was almost impossible to quench Roberts's thirst for personal glory, which nonetheless was satiated in abundance in the later course of the war.

Browne's column advanced into the Khyber Pass with two divisions, included in which were some of the Indian Army's most illustrious regiments, such as the Bengal Lancers and the Corps of Guides. Browne's was the first of the three columns to come into open combat with the enemy. At dawn on 21 November, Browne marched his men up the narrow track leading to the fort of Ali Musjid, the first Afghan position to come under attack. The deal struck with the Afridi Pashtuns of the Khyber held – the tribesmen joined in the fight that lasted throughout the day, until all resistance was silenced and the Afghans abandoned the fort under cover of darkness, leaving behind their guns, stores and camp equipment. The dagger-wielding Afridis chased after their fleeing brethren with the same enthusiasm they showed for plundering a British rearguard column. Browne moved quickly[34] on to Dakka and Jalalabad, which he occupied in December without resistance.

Following this action Lytton wrote a private letter to Cranbrook, in which he claimed the abandoned fort was littered with copies of proclamations by Sher Ali, calling on all Muslims to perform their sacred duty by driving the infidels from Afghanistan. Captured Afghan soldiers, according to the viceroy, carried the Koran, with the violent passages marked for daily study. Even if this allegation were true, it would be hard to comprehend Lytton's taking offence over the emir's belligerence. From the day the Chamberlain mission was sent back from the Khyber Pass, Sher Ali had become a king battling for his throne. One is reminded of Sir Olaf Caroe, Governor of the North-West Frontier Province in 1946, one of the most knowledgeable British administrators in India on the Pashtuns, whose classical poetry he translated. Caroe was once ingenuously asked why he thought the tribesmen, during the century of British rule on the Frontier, never relented from staging raids on British settled districts and attacking army columns. Caroe gave the obvious reply that if in 1940 the Germans had invaded England, he hoped that every man and woman would have resisted with the same ferocity the Pashtuns showed towards the British.

Roberts's column was by far the smallest of the three that comprised the invasion force, though geographically it was the closest to Kabul and thus the most worrisome for Sher Ali. While Browne was marching triumphantly past his first objective in the Khyber Pass, Roberts entered the Kurram Valley where, moving north, he was to encounter the first large concentration of Afghan troops. Winter was approaching and the nights were already bitterly cold. Roberts knew there was no time to be lost if he was to secure his objective before the snows closed in, halting further progress. This objective was the Peiwar Kotal Pass that gives access to the central plain of Afghanistan and Kabul, less than 90 miles away. Roberts marched his force up a single track road to Khapianga fort, which he found to be abandoned, and from there on to Kurram, with not a single Afghan soldier in sight. Local villagers told the British that the Afghans had got wind of the army's approach and had hastily fallen back on Peiwar Kotal.

Roberts received the news over lunch with the village headmen, where he also learned to eat Afghan-style. 'Knives and forks were evidently considered unnecessary . . . so I unhesitatingly took my first lesson in eating roast kid and pilau chicken without their aid.'[35]

But when an advanced contingent of Indian battalions reached the foot of the pass, expecting to quickly take the position, they found the ridge heavily defended by enemy troops and artillery. This was the moment Roberts's tactical skills came into play. By 29 November the remainder of the force was encamped below the towering 11,300-foot pass. Roberts contemplated the apparently impregnable position with 'a feeling very nearly akin to despair'.[36] His men were not gladdened by the visit of local villagers to the encampment, who assured them they were hastening to their destruction. Each day, Roberts went out on a reconnaissance of the route his men would have to follow to launch a frontal assault. Such a strategy would have been tantamount to suicide, but this is precisely the impression Roberts wanted to convey to the Afghan Turi[37] camel drivers and Muslim troops under his command. The general was certain some of these half-hearted men would send word to the Afghans that he was planning to storm their position. His suspicions turned out to be well founded, for in the assault on Peiwar Kotal some of the men of the 29th Punjab Infantry fired their muskets to warn the enemy of their approach. The culprits were later rounded up and Roberts later had the unpleasant duty of signing the death warrant of the sepoy who fired the first shot and handing down stiff prison terms for the others. 'The effect of these sentences', in his own words, 'was most salutary.'[38]

This act of treachery, however, did not upset Roberts's plan: the Afghans were successfully misled. The enemy guns opened up on the south side of the pass, showing that they were preparing to repel a frontal attack. Most of Roberts's men that day believed they would be embarking on a hara-kiri mission the next morning, but that night Roberts assembled his senior officers in his tent and disclosed his true plan of attack. This involved taking a turning column up around a parallel valley, with Roberts at its head and taking no animals or supplies with them, to give themselves the benefit of speed and surprise. Camp fires were kept burning in the valley and patrols were sent marching around loudly to distract the Afghans. Roberts led his men on the march all night on a 12-mile trek to reach a position at the Afghans' right flank. Just before daybreak, the artillery of the reserve troops left behind at camp opened a bombardment on the heights, which helped divert attention from the advance of Roberts's 5th Gurkhas and 72nd Highlanders.

At 6 a.m. on 2 December Roberts launched a savage attack on the Afghans, the Highlanders capturing two guns and winning a Victoria Cross within the space of an hour. The Afghans put up a desperate resistance at first, giving ground inch by inch at the point of the bayonet. Roberts swiftly changed tactics and opened a second turning operation, sending his column behind the enemy to cut off their line of retreat. When the Afghans realized what was afoot, the fighting withdrawal turned into a panic rout, the Afghans scattering in every direction in an attempt to escape being cut off. By midday the heights and plain below had been cleared of the enemy.

Peiwar Kotal marked a major victory and the final reversal of fortune for Sher Ali. For Roberts, it marked his first victory as a general at a cost of only 20 killed. It was always almost impossible to calculate enemy losses, for the Pashtuns would risk their own lives to drag their wounded or dead comrades from the battlefield to prevent them falling into Feringhee hands. But Afghan casualties were estimated at least in the hundreds. This was also the first major victory of the war. The Commander-in-Chief, India, General Sir Frederick Haines, as well as Queen Victoria, sent Roberts personal messages of congratulations. Roberts was knighted and he received the unanimous thanks of both Houses of Parliament.

While England and British India hailed a new hero, General Stewart moved out with the two divisions under his command. The tramp across the Baluchistan Desert was arduous: the troops had to negotiate three difficult passes, including the infamous Bolan, and in all, the distance this force had to cover was far in excess of the routes followed by Browne and Roberts. But with regard to engaging the enemy, the march on Kandahar 'had the character, for the most part, of a military promenade'.[39] Generals Stewart and Biddulph rode into Kandahar unopposed on 8 January 1879, where they learnt that the city's officials had fled in the direction of Herat when they were told of the army's approach. After a few days' rest, the Kandahar Field Force cavalry under Stewart pushed on to Kalat-i-Ghilzai, the fort which Nott had levelled to the ground on his homeward march in 1842. Biddulph went ahead towards Herat to occupy the town of Girishk, 78 miles west of Kandahar on the Helmand River. The little fortress here had been another battleground in the First Afghan War, where a small British garrison successfully withstood a nine-month siege by an overwhelming Afghan force until its relief by Nott's column.

By the end of January, the Khyber and Kurram passes had been swept clean of Afghan regular troops, Jalalabad and Kandahar were occupied by British forces and in less than three months of campaigning, the greater part of southern Afghanistan was in British hands. The affront to Chamberlain in the Khyber Pass had been avenged, the emir's standing army had been dispersed beyond all possibility of recovery and Lytton was rejoicing in the army's achievement. 'The rapid success of our military operations completely confirmed the calculations on which they had been based', the viceroy wrote to Cranbook.[40] Those with long memories would have given a shudder on recalling the parallels with the Army of the Indus's rapid successes of 1839, and the aftermath of that campaign.

In the royal apartments of Kabul's Bala Hissar, the emir contemplated the melancholy fate he had inherited from his father Dost Mohammed nearly 40 years before. Unlike his predecessor, Sher Ali was resolved not to submit to the British conquerors, to be packed off in disgrace to a life of exile in India. Yet his authority was collapsing and what few soldiers had remained loyal were rapidly deserting his standard. Browne at Jalalabad was the closest to Kabul of the three commanders of the Afghan Field Force. About this time, though there is no record of the exact date, Browne received a letter from Sher Ali, in which the emir announced his intention of abdicating in favour of his son Yakub Khan, the 'ill-starred wretch' he had imprisoned in 1874 for refusing to accept the designation of his now deceased brother, Abdullah Khan, as heir to the throne. In this letter, the emir gave out another piece of intelligence which gave those who had pushed for the invasion their vindication. Sher Ali told Browne he would travel to St Petersburg to expose his humiliation at the hands of the British, not only to the Russians, but to all of Europe.

Along with the handful of Russian agents who had remained behind in Kabul when war broke out, Sher Ali set off for Afghan Turkestan, a bitter and ailing man of fifty-five. He tried in vain to enlist Kaufmann's support in the war, never stopping to ask himself what the Russians had to gain by risking a conflict with Britain over that country's diplomatic dispute with Afghanistan. A packet of letters caught up with the emir at Mazar-i-Sharif, near Afghanistan's northern border, where ill health had forced him to rest. When he opened the letters, he realized there was no longer any room for hope: Kaufmann categorically refused to send troops to fight for the Afghan cause. Sher Ali was informed that he was welcome to join the general at Tashkent, but as for undertaking his onward journey to Russia, for the moment no authority for this visit had been granted by St Petersburg. This was the final blow which, combined with ill health and despair, put an end to Sher Ali's life on 21 February 1879.

Lytton in Calcutta, Cranbrook in London, in fact the entire host of panic-stricken Russophobes on both sides of the globe, could exult in their victory over the Afghan foe. But it was by no means a comfortable triumph, for whatever evidence was advanced to implicate Sher Ali in Russian plans to spread their influence in Afghanistan, it was equally evident that the emir was not an enemy of Britain. Like his father before him, Sher

Ali had made every effort to avert Russian adventurism in his kingdom. In the end, however, snubbed by a vainglorious Lytton and confronted with Stoletoff's threats, he had little choice but to acquiesce to the Russian mission.

It later came to light that ten days before the British invasion, a messenger from Kabul had tried to deliver another letter from the emir to Browne at Ali Musjid, in which he gave his consent to a temporary British Mission at Kabul. This letter never got through, as the messenger turned back when he heard that the fort had fallen. Sher Ali sent a second letter with a similar proposal to Cavagnari, who was with the Jalalabad garrison, and he rejected the offer on the grounds that it contained no apology for the Chamberlain incident in the Khyber Pass. It is almost impossible not to draw a parallel with Auckland's declaration of war, the Simla Manifesto, which was issued in October 1838 after the Shah of Persia had lifted the siege of Herat. Sher Ali was a ruler who bore no hostility towards the British. Now the accession of his son, Yakub Khan, to the throne opened the door to a negotiated settlement of the conflict.

It was in fact Yakub Khan who took the initiative by writing a letter to Cavagnari, informing him that the emir had died or, as he expressively put it, had 'hastened to the region of divine mercy'. Yakub also professed his desire to forge a bond of friendship with the 'illustrious British Government', which Lytton took as an offer to make peace. It was just as well that the new ruler was inclined to bury the hatchet, for Lytton's first instinct was to dismember Afghanistan into small, princely states that could be easily controlled. Kandahar would be a British protectorate, Herat relinquished to the Persians (notwithstanding that Britain had gone to war to prevent this happening), and Kabul would be placed in the hands of a pro-British friend of Cavagnari's. He held that all Afghan kings were an unscrupulous and untrustworthy lot, with strong Russian sympathies. Lytton's conditions for a cessation of hostilities were what he defined as the three main objects of the war: the punishment of Sher Ali (this condition had already been rendered redundant), the permanent 'improvement' of British India's frontiers, and British political dominion of the Pashtun tribes between the present frontier and the Oxus.

The viceroy urged Yakub to form a government quickly to negotiate a treaty that would meet British demands. To encourage fast action, on 12 April Browne's force moved from Jalalabad to Gandamak. By the end of the month some 5,000 men of the Khyber Field Force were massed on the spot where the remnants of the 44th of Foot had been wiped out on the retreat from Kabul. 'Here, four miles from the town of Gandamak, the bleached bones of Elphinstone's soldiers still covered the hills.'[41] Browne's column stood less than 70 miles east of Kabul and, as Yakub was only too aware, a short march from the capital. Unlike the previous war, this time the army was taking no chances of leaving its men exposed and with the risk of being cut off. A cavalry and two infantry regiments, some 1,600 men with sappers and miners, were holding Jalalabad, while another 6,300 troops stood ready at Jamrud, at the mouth of the Khyber Pass, forming one effective line of communication and supply. Roberts was placed in readiness to move on Kabul from Kurram, in case Yakub refused to submit to British demands. Lytton spoke the harsh language of the conqueror, but in reality he was anxious to recall the army to India at the earliest date. The last thing the Home Government wanted was to see Britain mired down in an expensive and protracted occupation of Afghanistan, such as had brought disaster in the first war.

Political considerations were also in play, with Disraeli facing anger at home over battlefield reverses in the Zulu War in South Africa, specifically the defeat at the Battle of Isandlwana, which cost 1,300 British casualties.

Lytton was prepared to restore Kandahar and eventually Jalalabad to Afghan sovereignty, and that was about the most generous of the entire list of terms. In spite of this ostensible gesture of goodwill, the conditions set down in the proposed treaty were decidedly harsh, lacking any trace of magnanimity in victory. Yakub Khan, in a more conciliatory tone, told Cavagnari in early April that he was prepared to accept the visit of a British envoy to Kabul. It was an extremely bold move on the emir's part, as many of his advisers remained violently opposed to any rapprochement with the British. Some believed there was still a chance of enlisting Russian support, and one concern Yakub expressed, rather prophetically, was that in such an environment no one could guarantee the safety of a British mission. Nevertheless, the end of the war was now in sight, though the army still had to contend with pockets of resistance, notably a Mohmand force assembling near the village of Dakka on the Afghan border.

The Mohmands, living in north-eastern Afghanistan and Pakistan's North-West Frontier Province, had always been the least trustworthy of the Pashtun tribes. Their homeland is divided by the Durand Line, the frontier demarcation that was forced on the Afghans in 1893. The emirs of Kabul relied on the Mohmands to stir up trouble for the British, and later the Pakistanis, whenever political circumstances required a diversionary ploy. A spirited charge by Sam Browne's Bengal Lancers dispersed the tribesmen who were threatening the village and that put an end to the last passage of arms of the war.

On 8 May, Yakub Khan, with a following of about 400 persons, reached the British camp at Gandamak to discuss the Government's peace terms. Lytton had telegraphed the Cabinet in London a draft treaty of 14 articles. The most substantive and potentially contentious of these concerned Britain's territorial demands, the placing of Afghanistan's foreign policy in British hands, and the acceptance of a permanent envoy in Kabul. The viceroy demanded the right to garrison Herat with British troops whenever this became necessary for frontier protection, but this was the one proposal the Cabinet saw fit to reject. Negotiations between Cavagnari and Yakub Khan began on 10 May and went on for a gruelling week until, after consultations with the Home Government, the final agreement was signed on 26 May 1879.

Leaving aside some of the less strategic points of the Treaty of Gandamak, Yakub Khan reluctantly ceded to British administration the Pishin and Sibi Valleys near Quetta, bringing British forces to within striking distance of Kandahar, and the Kurram and Khyber passes, which secured the army access to northern Afghanistan. The Government of India was given control of Afghanistan's foreign relations, which in essence prevented the emir from entering into agreements with any power regarded as hostile to British interests, viz. Russia. In return, Britain was committed to support Yakub against aggression and pay an annual subsidy of six lakhs (600,000) of rupees, roughly £60,000. Lastly but not least important, Yakub Khan agreed to accept a permanent British mission in Kabul, to the exclusion of any other power, thus enabling Britain to monitor Russian approaches and enforce Afghan compliance with Government policy.

In London, the treaty was hailed as a triumph for the Forward Policy, in line with Disraeli's wish for a 'scientific frontier' for Britain in Afghanistan. Browne's Khyber column was withdrawn, with the exception of two brigades, and orders were sent to the Kandahar column to prepare to return to India through the Bolan Pass. Roberts was told to hold on at Kurram, which had been transferred to British control and was to be administered by the Government of India. It now seemed that Afghanistan had become a British protectorate. But the treaty had thought of everything except the people with whom it was concerned.

Sir Louis Cavagnari received his knighthood for the diplomatic skills he displayed at Gandamak and he was awarded the additional honour of being appointed British Envoy to Kabul. There was no doubting Cavagnari's Frontier experience and his ability as a soldier and administrator. He served in the Oudh campaign during the Sepoy Mutiny and as Deputy Commissioner of Peshawar he distinguished himself in several expeditions against the Pashtuns. But Cavagnari held the Afghans in contempt and his arrogance generated a good deal of resentment amongst the chiefs in Kabul. Roberts had grave misgivings about this appointment. His personal view was that peace had been signed too quickly and that the army should have shown its strength by marching to Kabul and there dictated the terms of the treaty. This way, 'there would have been some assurance for its being adhered to. As it was, I could not help feeling there was none, and that the chances were against the Mission ever coming back.' Cavagnari showed no signs of sharing his forebodings and Roberts's heart sank when he wished him good-bye. 'When we had proceeded a few yards in our different directions, we both turned round, retraced our steps, shook hands once more, and parted forever.'[42]

Cavagnari entered Kabul on 24 July, where he took up residence in the Bala Hissar with a small staff consisting of a secretary, a medical officer and a military attaché in charge of an escort of 25 cavalry and 50 infantry of the Corps of Guides. Six weeks later they were all dead.

CHAPTER 7

Nothing but Misfortune and Disaster

On 5 September 1879, General Sir Frederick Roberts and his wife were enjoying a well-deserved rest in the delightful climate of Simla, the summer capital of the Raj. The party season was drawing to a close with the last of the gala dinners and balls, whilst the Government made preparations for the long trek down country to Calcutta. Within a few weeks, the bullock carts and carriages would embark on the more than 1,000-mile journey from the rhododendron- and pine-clad Himalayan hills to the steaming plains of Bengal. Lytton was enamoured of Simla, so much so that he chose this town for India's first exercise in town planning, with the design of the Viceregal Lodge and other grand neo-Gothic buildings.

On that day, around 1 a.m. in the morning, Roberts was awakened by his wife telling him that a telegraph man had been wandering about the house for some time. Roberts went downstairs and took the telegram up to his dressing room. It was an urgent message dated 4 September from Captain John Connoly, political officer at Ali Khel, near the Afghan border. Roberts hardly needed to open the envelope to foresee the bad news: 'The contents told me that my worst fears – fears I had hardly acknowledged to myself – had been only too realised.'[1] The telegram was from one of Cavagnari's secret agents in Kabul. The messenger had fled the city to send a report that on the previous day (3 September), the residency had come under attack by three Afghan regiments, later being joined by a portion of six other regiments, all heavily armed. The British garrison was last seen defending itself against the mob.

In his supreme arrogance, Cavagnari appeared to have had no inkling of trouble brewing outside the Bala Hissar. Given his long years of Frontier experience, he should have known better than to have taken for granted that the British mission would remain unchallenged for long. 'The presence in Kabul of the British agent who had negotiated the Treaty was in itself a constant affront to Afghan pride.'[2] As late as the end of August, Cavagnari was firing off optimistic telegrams to Simla, *à la* Macnaghten to Auckland in the days leading up to the Kabul uprising of November 1842. Cavagnari assigned little importance to the mutinous behaviour of the regiments that had marched from Herat to the capital, with bands playing, to demand payment of salaries in arrears.

The Herat regiments encamped two miles from Kabul, where they remained for four days. During this time they were harangued by their commanders, who urged the men to parade through the streets shouting abuse at Cavagnari and the Feringhees living in Kabul. One of the few witnesses of the massacre, Major Nakshband Khan of the Guides Cavalry, took the news of the soldiers' taunting and threats to the envoy. Cavagnari received this report with a mixture of cockiness and fatalism: 'Never fear, keep up your heart', he told the messenger. 'Dogs that bark do not bite.' To which the native officer replied, 'These dogs do bite, and there is real danger.'[3] Cavagnari was still in a flippant frame of mind: he assured the cavalry officer that if they were attacked, the Afghans could only kill the three or four Europeans at the residency, and that if this were to happen, their deaths would be well avenged. He was right on the second count, but wide of the mark on the first.

The Afghan troops were working themselves into a frenzy, and this was not helped by a cholera epidemic raging violently across Kabul, which killed some 150 of their men in a single day. Their hatred of the foreigners flared up when one of the ringleaders shouted for them to march on the residency to force Cavagnari to pay them their salary arrears. If the Afghan Government would not meet their demands, the emir's rich English allies should be made to pay. Several people sympathetic to the mission were turned back by the emir's sentries at the Bala Hissar, when they attempted to warn Cavagnari of an impending attack.

On the morning of 3 September, the mob approached the envoy's stables, where they proceeded to stone the guards and make off with the horses. It started off as an act of vandalism more than an organized assault on the residency. The mood turned ugly when several of the *sowars*[4] opened fire on what was still but a horde of disorganized looters. That was when some 200 of the men ran from the Bala Hissar to fetch their comrades and their arms from camp. In an hour's time, whilst confusion reigned in the residency, the Afghans were back in force. The serious fighting began at 8 a.m. that morning. The Afghan officers ordered the troops to attack the arsenal, which provided them with sufficient ammunition to mount a protracted siege. The store house was next to fall, then the treasury came under attack. By now much of the town had poured out of their homes to join the rioting and looting. An hour later, Cavagnari and his British comrades tried to regain the initiative. The envoy charged out of the residency door at the head of some 25 soldiers of the garrison. They made four sallies in all that day, and on each occasion the undisciplined Afghans were routed. But near midday flames were spotted dancing about part of the roof, which some eye witnesses believed was the work of Cavagnari's own men, in order to have less space to defend.

The battle went on until nine that night, until no more shots were heard from inside the building and the Afghans dispersed back to their camp. To the onlookers' astonishment, there was still some sporadic firing heard at the residency the next morning, but it soon became apparent that this came from the attackers moving in to finish off the wounded. Once they were assured that the entire garrison lay dead, all the bodies were flung into a pit dug for the purpose, though Cavagnari's corpse was never found and had probably been burnt in the house. The mission never stood a chance against the mob. Yakub Khan, who had 2,000 men under his command, could hear the cries and the shooting from his apartments only 250 yards away. The emir ignored Cavagnari's pleas for help, torn as he was between showing loyalty for his people's jihad and setting himself up as the target for British revenge. The Afghans fired from the cover of surrounding houses and even brought up a gun to pound the residency at close quarters. Too late Cavagnari sent out a message offering to pay the troops six months of their arrears.

Lytton was shattered by the news that reached him by telegraph at the Viceregal Lodge in Simla. He conveyed his desolation to Disraeli, neatly attributing the massacre to 'the hand of fate' instead of his own misguided decision to force the mission, and a weak one at that, on the Afghans. We have a hand-wringing Lytton lamenting the collapse of 'independent government in Afghanistan' as well as the 'obligation to occupy Kabul', a curious forward-looking concern about evacuating the city without bringing 'renewed disaster to Yakub Khan' and the need 'to undertake the virtual administration of the country'.[5] The task of reconciling these contradictory utterances would be left to the army to resolve. In a word, the Government had no choice but to launch another and probably costlier campaign in Afghanistan, with Kabul, the seat of the insurrection, as the objective. Roberts was chosen to lead this punitive expedition to the capital and occupy the city. Browne's Khyber force had been broken up, while Stewart was in Kandahar with a token detachment, the bulk of his column having left on their return march to India. Roberts was ordered to proceed at once to Kurram and resume his command, a job he took on with relish: the little general sensed that his newly-named Kabul Field Force was destined for glory.

Roberts spent the hours remaining before departing Simla locked in strategy talks with Commander-in-Chief Sir Francis Haines, to determine the composition of the force to be raised for the advance on Kabul. As soon as the news from Kabul was confirmed, Stewart was requested to recall his troops to Kandahar, where the emir's officials, lacking any guidance from Kabul, had no choice but to place themselves under the general's authority. Browne at the Khyber was speedily reinforced in order to form an effective blockade to any attempt by the Afghans to harass British India from that vulnerable route. Lytton sent a message to Yakub Khan to inform him that the army was on its way to Kabul. The viceroy at this confused juncture believed the emir was in need of British protection, whilst Roberts took a more hard-nosed view of Yakub's true intentions. More troops had to be moved up to Kurram to enable Roberts to march without compromising his base of operations.

The general insisted on having seasoned battlefield commanders to lead his two infantry brigades. His wish was fulfilled with the appointment of Sir Herbert Macpherson, a VC holder who had served in the relief of Lucknow during the Mutiny as well as three major Frontier campaigns, and Sir Thomas Baker, a former aide-de-camp to the queen and military secretary to Lytton, and a veteran of African campaigns. Roberts then paid a farewell visit to Lytton, whom he found, as might be expected, 'in a state of deep distress and depression'.[6] Lytton had formed a genuine bond of friendship with Cavagnari, whose murder left the viceroy grief-stricken. Beyond that tragedy Lytton had more pressing matters to consider: a foreign policy in tatters and the North-West Frontier of India unsecured.

Roberts departed Simla on 6 September, thinking to deploy his 7,000 men and 22 guns in a swift advance on Kabul. It was a challenging task, for 'his line of communication and supply stretched back two hundred miles to India, and he was without adequate transport animals or experienced transport officers'.[7] The success of the operation depended on speed. Kabul had to be reached before Yakub Khan – in Roberts's eyes a traitor who had stirred up his people against the British – was able to mount an organised resistance to his advancing force. Roberts had a keen ability to second-guess his foes. He reasoned that the Afghans would not oppose his column until it reached the Shutargardan Pass, a critical objective he would need to take before an early fall of snow rendered it impassable. Roberts needed to be in control of the open countryside, for once past the Shutargardan, his men would be thrown on the resources of the land or supplies forwarded by the line of the Khyber, in a region swarming with hostile tribesmen. Yakub Khan knew that when the British had cleared the Shutargardan, they would be within striking distance of Kabul.

Roberts sensed that it was here at the pass where he would come up against Afghan resistance, and his intuition did not let him down. A party of some 2,000 of Yakub's armed followers blocked their passage at the narrowest part of the defile. No sooner had Roberts's column come up against the enemy, a volley of shots exploded in the canyon walls. A large party that had been lying in ambush spotted Roberts at the head of the column and directed their fire at him. He luckily escaped unscathed, though the chief medical officer riding at his side fell severely wounded with a bullet in his cheek. Roberts called in the 92nd Highlanders and a party of dismounted Lancers, who quickly and with comparatively little effort scattered the Afghans to the hills.

Roberts stood on top of the Shutargardan Pass on the evening of 27 September, surveying the vast Afghan plain below, when a messenger rode up bearing a message from the emir. In it, Yakub Khan tried to dispel Roberts's unease that the Khyber column might be met with opposition. Yakub, obviously fearful of retribution for the massacre at Kabul having taken place on his watch and his failure to aid Cavagnari, professed nothing but loyalty to the British Government. To Roberts's annoyance, Yakub proposed a personal meeting at Kushi, a spot 14 miles beyond the pass, where General Baker was stockpiling an extensive depot to back up the army on its advance to Kabul. Even more annoying was that Yakub, without waiting for a reply from Roberts, had already reached the British encampment uninvited, attended by his son, his father-in-law, the commander-in-chief of the Afghan Army and an escort of 200 men. Roberts was ever mistrustful of the Afghans and had the Shutargardan garrisoned with a mountain gun battery and two infantry brigades before setting off for Kushi. The general was not favourably impressed by the emir, whom he described as 'an insignificant looking man . . . with a receding forehead, a conical-shaped head and no chin to speak of'.⁸ Roberts perceived him as a man utterly lacking in force of character to govern a warlike and turbulent people like the Afghans. 'I felt that his appearance tallied exactly with the double-dealing that had been imputed to him.'⁹

This first encounter with the emir, from Roberts's standpoint, was nothing more than a tiresome formality. Yakub Khan tried to ingratiate himself with the general by asking for time to restore order amongst his troops, and find and punish those responsible for the attack on the residency. Roberts listened patiently until the emir had finished his flamboyant discourse, and then dismissively replied that his orders, as well as his determination, were to press on to Kabul with all possible speed. Yakub stood back visibly alarmed by Roberts's unforgiving tenacity. The general's only concession was to allow the emir to evacuate all women and children from the city if an attack should prove necessary. Roberts had not forgotten the carnage that took place in the relief of Delhi in the Mutiny, when British troops had been forced to fight their way at bayonet point through narrow streets and crowded bazaars. Unwelcome a guest as Yakub was, Robert agreed to offer him hospitality at his camp, firstly to keep informed of his movements and also to prevent him becoming a rallying point for resistance in the forthcoming march on Kabul. Roberts was not of a mind to waste time: that same day he sent a proclamation to Kabul, warning non-combatants to clear out of the city, and announcing that anyone found bearing arms would be treated as an enemy. The entire force, with Yakub Khan and his party in tow, moved off on 2 October.

The British were in an advantageous position militarily. In addition to Roberts's large force, a column of upwards of 16,000 troops under General Sir Robert Bright was on its way up the Khyber, and the large number of men under Stewart had reoccupied Kandahar and Khelat-i-Ghilzai, from where they were threatening Ghazni. Roberts was four days into the march when, on 6 October, his scouts galloped up to the main column with reports of a great body of tribesmen advancing from Kabul. By now the Kabul Field Force had swelled to more than 6,600 fighting men, with another 6,000 camp followers and some 3,500 baggage animals making up a train of men and beasts that stretched for miles across the plain. Roberts had no way of knowing what resistance he would encounter on the road, hence the column carried a fortnight's supplies, but oddly enough with two months' worth of tea and sugar, the two commodities most readily obtainable from local sources.

Roberts was now in sight of the last obstacle before Kabul, a crescent range of hills just before the village of Charasia, about seven miles south of the city. The hill was bisected by a small pass close to the Logar River, whose waters the troops had to ford whilst simultaneously driving off attacking bands of tribesmen. Once the force was encamped, native spies began arriving with reports of regular troops as well as citizens of Kabul arming to repel the British advance. The last stretch of road was impassable for guns, so Roberts himself went out with an escort of cavalry to examine the terrain. The growing daylight revealed large numbers of Afghan regulars crowning the hills on both sides of the pass, who coolly went about placing their guns on the heights. The next report brought into camp told of the road to the rear being blocked and that Macpherson's brigade that was still behind at Safed Sang was in for a rough time. The transport was meant to be sent back to collect the force and stores later that day, but there was no way of knowing if the wagons would be able to break through and return before dark with the badly-needed reinforcements.

The situation was one of great anxiety, as Roberts had feared. The emir had used a system of messengers to keep the Afghan commanders in Kabul fully informed of British troop positions and strength. Shortly before midday, General Baker's men were suddenly and hotly engaged by Afghan regulars and a swarm of armed tribesmen, in all some 13 regular regiments of infantry, 20 field guns and several thousand tribesmen. The Afghans could rain down a hail of bullets and shells at their leisure from their commanding position above the road. Roberts was sitting with only two-thirds of his force on the flat ground beneath the imposing hills. He had to act fast and accept that Macpherson might not make it through in time to help him. He quickly ordered one company of the 72nd Highlanders to take the almost inaccessible ridge, which was alive with Afghan fighters. It was crucial to push back the enemy before they could bring up reinforcements. There was no respite for the troops that morning and it took nearly three hours of fierce combat for the Highlanders, supported by Gurkhas, to clear the heights. As they did so, often at bayonet point, the Afghans continued to rally and contest every inch of ground on their retreat.

Shortly before sunset, Roberts received a heliograph[10] message, 'a most satisfactory one, to the effect that the whole of the enemy's position was in our possession, and that our victory was complete'.[11] This was the first time a heliograph was used in battle, a technological advance that left Roberts decidedly unimpressed, for instead of being where he was happiest, in the thick of the fighting, he was now expected to direct the action from a distance. British losses amounted to 86 killed and wounded. 'The enemy's loss is unknown, but it must have been considerable. They fled in great confusion, and lost two standards.'[12] Throughout the day, the emir and his senior commanders remained seated on a knoll in the camp watching the progress of the battle. Once Roberts was assured that the day was his, he sent an aide to Yakub to convey the joyful news of his success. 'It was, without a doubt, a trying moment for him, and a terrible disappointment after the plans which I subsequently ascertained he and his adherents at Kabul had carefully laid for our annihilation.'[13]

With news of Ghilzai tribesmen gathering for an attack on the Shutargardan and Afghan regular army reinforcements moving up to Kabul from Ghazni, Roberts was compelled to push on despite the troops' clamour for rest after their hard-fought battle. The men rallied to their commander's call and a rapid advance soon brought them to the town of Beni Hissar, only two miles south of the imposing Bala Hissar citadel looming on the horizon, and which to Roberts's joy was found to have been evacuated. He was informed that the Afghans were hastening north towards Bamiyan and Kohistan, so he dispatched a cavalry brigade to cut off the enemy's retreat. The Afghan troops remaining in Kabul had occupied a ridge known as the Asmai Heights north-west of Kabul, above the Sherpur cantonment, near the spot where the Army of the Indus had built their quarters.

Roberts decided to launch an immediate attack. General Baker was ordered to swarm the Afghan position, by first bringing in guns for a barrage to soften up the enemy. This proved ineffectual and the action got bogged down in a stalemate until, on the following morning of 9 October, it was discovered that the Afghan troops had decamped during the night. Roberts was delighted to be in charge in Kabul, the place he 'had heard so much about from my boyhood, and I so often wished to see'. He had emerged as the indisputable hero in this next stage of Britain's Second Afghan war and, as he stated, his success had been complete. 'All opposition had been overcome, Kabul was at our mercy, the Emir was in my camp ready to agree with whatever I might propose, and it had all been done with extraordinarily little loss to ourselves.'[14]

The first stage of the invasion had ended with Roberts as master of Kabul, a fact that was driven home to the citizenry three days later in a full dress parade, on 12 October, with band playing, to mark the army's formal occupation of the Bala Hissar. Roberts could not have been more pleased, for on that morning Yakub Khan marched into his camp with only two followers to announce his abdication. He complained to Roberts that his life had been a miserable one, and that he would rather be employed as a grass-cutter in the English camp than Emir of Afghanistan. In the tradition of deposed Afghan rulers, Roberts granted Yakub Khan permission to shelter in his camp until such time that he could be dispatched into exile in India. He departed Kabul on 1 December, never to return. The day following the emir's abdication, 13 October, the Kabul Field Force marched through the streets of Kabul, now officially in British hands.

While Roberts consolidated his hold on Kabul, the Khyber was being reinforced with a large contingent of troops to ensure this vital lifeline to Kabul remained open once the winter snows had closed the Kurram route over the Shutargardan. Browne was out of the picture: he had been given temporary command of the Lahore Division, a posting not to the liking of this fighting general, and from which he resigned a few months later. Browne's removal from the field came as no surprise – Lytton had an intense dislike of the coarse-spoken general and never allowed him to exercise the political power with which he had been invested. Major General Sir Robert Onesiphorus Bright, sporting a set of mutton chop whiskers nearly as flamboyant as his name, was in charge of the Khyber. General Stewart had quickly reoccupied Kandahar and recalled his troops, just as Nott had done in the previous war, but he was not destined to set foot in Kabul until May of the following year.

The people of Kabul assumed that once Roberts had taken his revenge for Cavagnari's murder, he and his army would depart the city in order to reach India before the winter snows blocked the passes. But they were mistaken. This time the British had not come to place a feeble and detested puppet king on the throne – it would instead be occupied by the most revered general of the British Raj, who in effect took on the mantle of the new Emir of Afghanistan. The country was placed under British rule, Roberts's force having come not as the ill-fated Army of the Indus, but as an incarnation of Pollock's Army of Retribution. This was entirely in keeping with official instructions given to Roberts in Simla. 'It is not justice in the ordinary sense, but retribution that you have to administer on reaching Kabul', Lytton had confided to him at a farewell meeting. 'Your objective should be to strike terror, and to strike it swiftly and deeply.'[15]

At midday on 11 October, Roberts and his staff, accompanied by a large retinue of Afghan dignitaries, entered the Bala Hissar. The British flag was raised, the bands played the national anthem and a salute of 31 guns was fired. Roberts addressed the multitude of onlookers in his typical unadorned style: Kabul deserved to be destroyed and its very name blotted out; punishment inflicted on those implicated in the residency massacre would be felt and remembered; anyone found bearing arms was liable to punishment by death; buildings that interfered with the military occupation of the Bala Hissar would be levelled; and a heavy fine was to be imposed on the inhabitants of Kabul, which was placed under martial law. Roberts the de facto emir appointed Major General James Hill to be Governor of Kabul and a military court was set up to try those accused of having taken part in the attack on the Cavagnari mission. In all, 87 persons were 'officially' hanged for complicity in the massacre or disobedience to Roberts's proclamation. The actual number of executions was probably close to 100.

In compliance with the great act of retribution demanded by Lytton, Roberts had the interior of the Bala Hissar gutted, leaving only the outer walls standing, whilst his engineers completed work on the Sherpur cantonment, which from 1 November was to be the army's winter headquarters.[16] The Bala Hissar might have seemed the more obvious place to quarter the garrison, however it was dangerously close to a large magazine that had escaped destruction and the citadel was in fact not spacious enough to accommodate all the troops. Roberts had no intention of scattering his army about Kabul, hence his decision to move into the fortified cantonment of Sherpur, which had been built by the late Sher Ali as winter quarters for his regular troops.

The round of summary executions provoked an outcry of indignation in the British press and from Gladstone's Liberal opposition, so that Roberts was called upon to give an explanation of his conduct. In a statement drafted by Roberts and read to both Houses of Parliament by Secretary of State for India Viscount Cranbrook, Little Bobs stood firm in defence of his ruthless handling of the Afghans. One can build a case for punishing those who had taken part in the attack on the mission, though it is less easy to justify a full-scale invasion of Afghanistan to avenge that crime. The London *Daily News* correspondent who accompanied the Kabul Field Force makes clear that some of these men were executed merely for their opposition to this invasion. 'The Afghan Army, or such of it as exists, must see that we were thoroughly in earnest in threatening with death all who chose to appear as rebels against the then Emir, in whose name we were advancing.'[17] It was the person who had called for rough justice to be imposed on the Afghans that presented Roberts with his reward. Lytton appointed him supreme commander of all the forces in eastern Afghanistan, comprised of 20,000 men in two divisions.

With the onset of winter, the British began to prepare their quarters with stocks of provisions and forage. On 1 November, the day the army moved into the Sherpur cantonment, Macpherson left Kabul with a brigade of 1,800 men to link up with General Bright's troops advancing from the Khyber. The main priority was to maintain a route for wheeled carriages for the army's eventual return to India. Roberts was eager to avoid the difficult Khoord Kabul Pass, the scene of the wholesale massacre of Elphinstone's column in 1842. The Afghan commander-in-chief, General Daoud Shah, with some irony informed Roberts that a much easier route lay just to the east of the dreaded pass.

In Kabul, the Afghans began to put aside their internal quarrels in their common hatred of the Feringhee. The desperate fighting that took place in December proved that they were in earnest. One of the Afghans' great tactical advantages lies in their unpredictability as an enemy. 'There seems to be no relation between opportunity and the period of the inevitable outburst ... Perhaps the rising ripened faster in 1879 than in 1841 because in the former period no Macnaghten fomented intrigues and scattered gold.'[18]

In the final weeks of 1879, the daily assemblage at Kabul's vast, blue-domed Pul-e Khishti Mosque was swelling in numbers, passions were inflamed, and the faithful coming from evening prayers dispersed through the streets with growing anger, which no one in the British garrison could account for. Reports were circulating in the cantonment of an impending uprising in outlying parts, beyond the range of British guns. The situation in

Kohistan was tense, the Logar Valley was on the verge of rebellion, the Maidan and Wardak districts to the south-west were said to be in open revolt, with the former artillery commander and veteran of the Khyber Pass action Mohammed Jan leading the insurrection. One of his first acts was to proclaim Yakub Khan's eldest son, Musa Khan, emir. The boy's appointment was not intended to be more than a symbolic gesture of solidarity with his deposed father and never took hold. Intelligence was scanty and what few accounts filtered through to Roberts's headquarters at Sherpur arrived long after events had taken place. But all the reports concurred in that a 90-year-old firebrand fanatic from Ghazni, Mullah Mushk-i-Alam, was preaching jihad against the British oppressors from every mosque in Afghanistan, and that the populace was responding in droves.

The first stirrings of trouble had come as early as October, when the Shutargardan Pass garrison was twice stormed by a tribal lashkar,[19] which was easily beaten off with little loss to the British. But there was also an attempt to take the British position at Ali Khel to the south, and then in quick succession another attack on the Shutargardan. This latter position was abandoned when General Bright's force cleared the Khyber route to Kabul, and the troops were deployed to the capital and Ali Khel. The storm broke in December, in the wake of Yakub Khan's departure from Kabul with his ministers. Roberts was never in doubt as to what had raised to a high pitch the Afghans' natural antipathy towards a foreign invader. The infidels' occupation of Sher Ali's cantonment, the capture of the Afghan Army's munitions, the destruction of the historic Bala Hissar and the dispatch to India of the emir – all provocations in abundance to fuel an insurrection.

The Afghans were massing in three columns in the countryside, from the north, south and west, aiming to converge on Kabul, capture the city with the help of the inhabitants, and lay siege to the Sherpur cantonment. It was evident that the advancing lashkars would have to be met and scattered before they could join forces for the march on the capital. Roberts had only an imperfect idea of the enemy's whereabouts and fighting strength.

But that there was serious trouble ahead was plain enough. Before the Afghans had a chance to cut the lines, he sent an urgent telegraph to Bright at Jalalabad, requesting him to dispatch the Corps of Guides to reinforce his garrison. Macpherson marched with a column to prevent the lashkar coming from the north linking up with the one moving up from the west. Baker was instructed to place his troops across a line that would block the Afghan retreat, once they were beaten back by Macpherson.

The plan was for Baker to head south and throw his troops across the Ghazni road, whilst Macpherson, moving in sympathy, was to engage the enemy and cut off his northern escape route to Bamiyan. The two generals marched from Sherpur on 8 December, expecting to encounter a lashkar of around 5,000 men. They later discovered that Mohammed Jan had mustered twice that number. Once the Afghans found themselves encircled, it was reasoned, they would have no choice but to put up a stout resistance. Roberts was confident of defeating these hill fighters, who are unmatched in guerrilla warfare, in open combat. But things turned out differently. Mohammed Jan evaded the trap by outmanoeuvring the British columns, to install his force on the heights south of the Bala Hissar. As a result of this unexpected development, reported a journalist on the scene, 'The whole country seemed covered with masses of armed tribesmen, and on every low hill banners were flying.'[20] Macpherson turned his mountain guns on the Afghans and a few well-aimed shells managed to break their advance from the crests. The general brought up 13 companies of Gurkhas, Sikhs and British infantry, at the sight of which the tribesmen abandoned their hilltop positions. Macpherson set out to drive the remaining lashkars south towards Mardan, where Baker anxiously awaited with five troops of cavalry, 950 infantry and four mountain guns. There was good reason to believe that Mohammed Jan was at a position halfway between Macpherson's and Baker's troops.

Roberts's strategy would work if the generals were able to force the Afghans into a pincer between the two British columns. But they never came close to achieving their objective and the operation ended in disaster when Brigadier General William Massy, starting from Sherpur with a cavalry unit to catch up with Macpherson, advanced too far and found himself surrounded by several thousands of Afghans waiting across the Ghazni road, drums beating and green standards waving. Far from retreating supinely onto the muzzles of Baker's guns, Mohammed Jan was positioned inside the British line and facing towards Kabul. Only Massy's lancers and four guns stood between the Afghans and the depleted garrison at Sherpur. Of the options open to him – a slow fighting retreat or a frontal attack – Massy made the rasher decision, with calamitous results. The guns had to be drawn back after each fusillade under the weight of the enemy's overwhelming force.

Roberts was no longer prepared to sit it out at Sherpur, receiving a regular flow of distressing reports from the battlefield. He rode from the cantonment to Massy's line and once there he acted with decision. Massy was ordered to retire slowly and watch for an opportunity to extricate the guns under cover of a cavalry charge. Roberts instructed Macpherson to hurry to Massy's assistance, and then told his acting commander at Sherpur, General Hugh Gough, to dispatch a detachment of Highlanders to hold the Kabul River gorge at all costs. Massy gave the order to attack: a troop of 200 British cavalry charged at full gallop with lances in the engage position, straight into the heart of 10,000 Afghans. The horses kicked up such a cloud of dust that it was several minutes before Roberts and his aides contemplated the terrible spectacle of riderless mounts followed by broken groups of troopers emerging from the dust. Astonishingly, only 18 men had been slain in the charge, but their four guns had been lost. The 72nd Highlanders arrived in time to save the day, by checking the advancing tide of tribesmen. On the morning of 11 December, a bewildered Baker, finding no Afghans on the horizon, marched from Maidan, and after some sporadic skirmishing his column reached the cantonment the following day. In his report, Roberts laid on Massy at least part of the blame for the disaster. Massy was recalled to India but like Elphinstone in the previous war, he had influential friends at the Horse Guards, who had him reappointed to a brigade.

Roberts spent the next few days of comparative calm in strengthening the defences of Sherpur, whose fortifications were never totally to his liking. The perimeter was too large, even for a force of 7,000 men to defend. The primary concern was to hold the Bemaru ridge that dominated the cantonment. Roberts recalled Shelton's failed attempt to capture the ridge from which the Afghans effectively harassed the British cantonment in the previous war: 'To have given up any part of it would have been to repeat the mistake which proved so disastrous to Elphinstone's army in 1841.'[21] So long as the garrison could hold Bemaru, it was safe from attack from the north. Roberts had ordered six towers to be constructed on the ridge and shelter trenches and gun pits made at points where infantry and artillery could be used with the greatest advantage. This his remarkable corps of sappers achieved in just a few days. But the attack didn't come from the north.

Roberts took the disagreeable precaution of placing General Daoud, the former commander-in-chief, under arrest and later had him deported to India in a most gentlemanly fashion. Despite Roberts's fondness for the man, he sensed there was trouble brewing in Kabul and there was no taking chances of him turning coat and going over to the rebel side. Daoud harboured no ill feelings towards Roberts and the general was reported to be much enjoying his stay in Calcutta, where he became a frequent visitor to the city's zoo.

By 21 December, there were unmistakable signs of an attack at hand. Several posts to the east of the cantonment had been occupied by the Afghans and scaling ladders were being assembled, whilst in the mosques, the mullahs were making frantic appeals to exterminate the infidels. The aged Mullah Mushk-i-Alam, the chief religious instigator, was doing all in his power to fan the flames of revolt. To Roberts, 'This looked like business.'[22] The following night, three days before Christmas, the cold winter stillness was shattered by war cries from outside the perimeter walls. It was the last day of Muharram, the first month of the Islamic calendar, in which fighting is prohibited. Just before dawn, the flames of a beacon fire shot into the air on the topmost crag of the Asmai ridge, south-west of the cantonment, and firing began from the hills. This was the signal for the assault to begin.

The enemy advanced en masse in a lashkar of up to 100,000, according to estimates from Roberts's spies. Thousands of screaming, banner-waving tribesmen came streaming down from the hills and just when the fighting was at its hottest, the general experienced one of those amusing incidents of war. While Roberts was on the battlements directing the defence, his elderly Muslim servant came up to whisper in his ear that his bath was ready. At one point Mohammed Jan, apparently in all seriousness, sent a message through to Roberts suggesting the British evacuate Afghanistan, reinstate Yakub Khan as emir and leave behind two high-ranking officers to guarantee fulfilment of these terms. The Afghan general might have seen himself as an incarnation of Akbar Khan, but Roberts was no Elphinstone and took no notice of the overture. He had a battle to direct, the fighting of which raged for three hours throughout the morning. The Afghans hurled themselves at the south-eastern wall with fanatical zeal, leaving heaps of dead below the ramparts. They continued to throw wave after wave of their men at the perimeter, with total disregard for the enormous number of casualties they were taking, until Roberts ordered the 5th Punjab Cavalry with four field artillery guns to launch a counterattack.

The last thing the Afghans had expected was for the British garrison to go on the offensive. A spirited barrage from the guns drove them back in disarray. By 1 p.m., the enemy were in full flight. This was the Cavalry's opportunity: Massy's troopers followed in pursuit of the Afghans and before nightfall, the area around Sherpur was cleared of the enemy. Roberts's victory had been achieved with the loss of 18 men, whilst Afghans killed exceeded 3,000. Little Bobs then ordered forts to be erected on the Asmai heights and other raised positions around the cantonment (including one to the south called Fort Roberts) so that by New Year's Day, 1880, he could proclaim that Sherpur had been made safe and the outlook for Afghanistan was 'fairly satisfactory'.

The country appeared to be quiet, and General Stewart was planning to move his force towards Ghazni as part of a scheme to pacify the region and enable an early withdrawal from northern Afghanistan, once the Indian frontier had been made secure. As Roberts correctly surmised, like every would-be conqueror of this turbulent land, the British were obliged to face the eternal conundrum: 'What was to be done with Afghanistan now we had got it?'[23] Roberts himself was convinced the only way to ensure sustainable tranquillity was for the country to be broken up, 'disintegrated' in his own words. He believed that Afghanistan was ungovernable under a single sovereign. This was a radical proposal, but by no means an unreasonable one. Had this bold step been taken, Afghanistan might have developed into a workable federation. After all, Afghanistan had only existed as a unified state for a few decades, since the early nineteenth century. Dost Mohammed was the first ruler of that era to attempt to build a nation state, but it took him 21 years to establish his supremacy over Afghan Turkestan, Kandahar and finally Herat, where he died in the enterprise. The idea of a break-up would have under normal circumstances gained support from Lytton, but he had now been relegated to the role of lame-duck viceroy. Britain was heading towards a general election and the first warning notes of a change of Government, and with it the inevitable appointment of a new viceroy, were sounded in India in early January.

Britain at this time exercised effective military control of Kabul and Kandahar. The districts adjoining these two cities were also under British command. But as Lytton himself bitterly acknowledged, after two years of warfare and occupation, 'the range of our effective administration or influence went no further, so that the country at large was without a government'[24] – it was ever thus. The exception was Herat, where Yakub Khan's brother, Ayub Khan, had managed to maintain himself in power, though the man who would later defeat the British in the war's worst military disaster was hardly an acceptable candidate for emir.

Military and political operations in Afghanistan had ground to a standstill, a situation that brought the Conservative Government at home under heavy criticism. The massacre in January 1879 of 1,300 British troops at Isandlwana in the Anglo-Zulu War, and the assassination of Cavagnari and his staff at Kabul in the same year, coupled with the stalemate in Afghanistan, exposed Disraeli's imperial policy as a triumph of style over substance. The lack of any real achievement in South Africa and Afghanistan cost Disraeli the April 1880 election, ushering in Gladstone's second Liberal ministry.

All efforts were now focused on how to extricate the army from Afghanistan without compromising British India's security. The answer had to lie in a strong leader who was friendly to Britain, yet who in the eyes of his people remained independent of British dominion. Lytton transmitted the need to find a way out of this 'rat-trap', as he put it, to Sir Lepel Griffin, in preparation for the Chief Secretary of the Punjab's forthcoming visit to Kabul. The viceroy outlined a four-point policy directive aimed at bringing to a close direct British intervention in Afghanistan. Lytton ruled out the restoration of Yakub Khan and the proposed severance of western from north-western Afghanistan. There was to be no annexation or permanent occupation of the latter and, lastly, the Government should be willing to recognize any ruler chosen by the Afghans (except the ex-emir) to restore order and allow for the evacuation of British troops.

Then in early January, as if delivered by the hand of Providence, the man who would go down in history as the 'Iron Emir' came on the scene. Stern, heavily-bearded and with the charisma of a saint, Abdur Rahman unexpectedly made his appearance at Balkh, near Afghanistan's Oxus frontier. He was Dost Mohammed's grandson, and had taken refuge first in Bokhara in 1867 and three years later in Tashkent under Russian protection, when he lost his bid to oust Sher Ali in a civil war for the succession. The news of Abdur Rahman's return reached the British through a Reuters dispatch on 7 January. Here at last was Britain's man, 'who welded the loose congeries of turbulent tribes into a nation and ruled with a rod of steel over a united Afghanistan'.[25]

Abdur Rahman, a lone horseman on the plain, was greeted with an outpouring of frenzy usually reserved for the coming of a holy prophet. An army of 12,000 horsemen and infantry, which he thought had been drawn up on the plain to oppose his return, parted like the waters of the Red Sea at his approach. Whilst Abdur Rahman was inspecting the artillery, a messenger bearing a letter from Lepel Griffin threw himself at his master's feet. Recently appointed Lytton's political agent in Kabul, Griffin was the man destined to elevate Abdur Rahman to the throne of Kabul. He assured the burly Afghan that he would benefit much more from relations with the British than with Russia, that there was absolutely no intention of annexing Afghanistan, and that Britain's chief desire was to see a strong and friendly emir established at Kabul. The astute and understandably suspicious Abdur Rahman drafted a vague reply, explaining that his only intention in quitting Russia was to help his country.

The flipside of Abdur Rahman's wariness was the British Government's natural scepticism: he was by far the strongest candidate for leadership, yet his true intentions remained an enigma. In a worrisome harangue addressed to the chiefs of Kohistan, the claimant to the throne announced his readiness to march on Kabul to redeem Afghan honour. Paradoxically, he also stated in his letter his wish to be at peace with the British, if they would accede to his representations. But at the same time, he could not overlook his debt to Russia, whose salt he had eaten. Therefore Abdur Rahman wished Afghanistan to be placed under the protection of both European powers. The Government was never going to consider a power-sharing deal with Russia, though had the logistics of such an arrangement been satisfactorily worked out, it might have put an end to the sole cause of turmoil and bloodshed in that country. Hence Griffin and Lytton were thrown into a quandary on how to draw the Afghan chief away from the Russians and into the British camp, whilst also persuading the Ghilzais of Ghazni, who favoured Yakub Khan, that Abdur Rahman was their man.

Given the speed at which the tribes of Badakhshan and Afghan Turkestan were gathering around Abdur Rahman's standard, he might soon be in a position to dictate terms to the British, instead of accepting them from the Government. Lytton laid these concerns before Cranbrook, his fears reinforced by report of some fierce attacks having been launched in mid-April on Stewart's division at Ghazni and Colonel Frederick Jenkins's brigade at Charasia, on their march from Kandahar to Kabul. Roberts sent Macpherson with 962 infantry and four mountain guns, followed by two more guns and a troop of the 3rd Punjab Cavalry to relieve the besieged columns, and the enemy was soundly defeated in both places. On 28 April, the Kandahar and Kabul contingents joined forces and Stewart arrived at the capital on 1 May, bearing some unwelcome tidings for Roberts. As the most senior officer in Afghanistan, Stewart had come to assume overall command of the Kabul Field Force, which now numbered nearly 14,000 men and 38 guns. Another 15,000 troops and 30 guns stood ready on the Khyber line under the command of Major General Bright. As Roberts confessed, Stewart's arrival 'was altogether not a happy day for me'.[26]

On 2 May, news reached Kabul that Disraeli's ministry had come to an end, that Lytton had resigned and was to be succeeded by the Marquis of Ripon, and that the Marquis of Hartington had become secretary of state for India. 'I dreaded that a change of Government might mean a reversal of the policy which I believed to be the best for the security of our position in India', wrote Roberts.[27] It deeply pained him to be sidelined, and in a moment of despair he asked to be sent back to India, alleging poor health. Fortunately his request was turned down, for Little Bobs's greatest day of glory lay less than four months away.

Lytton sent a conciliatory letter to Abdur Rahman, promising to withdraw all British troops by October 1880 and inviting him to proceed to Kabul to take rightful possession of his throne. Abdur was not completely won over by Lytton's offer. He wanted to know what would be the status of Kandahar once the British had left, and for that matter what would be the boundaries of his kingdom and, of course, whether the British intended to leave behind another European representative at Kabul. Lastly, he asked Lytton to spell out what benefits would be conferred on him as emir and on his country. These issues were dropped in the laps of the new viceroy and secretary of state for India of Gladstone's ministry. George Frederick Robinson, afterwards Lord Ripon, was a wealthy Yorkshireman and the son of one of Britain's most ineffective of prime ministers.[28] Ripon's flowing white beard and steely countenance bestowed on him the appearance of a patriarch of the Orthodox Church, instead of the Roman Catholic convert he had become. In prosecuting the Afghan war, Ripon did not disappoint Roberts by taking a soft line. On the contrary, his Afghan policy brought him a triumph in his first year as viceroy, though 'he was not of any brilliancy or capacity' in his attempts at imposing liberal civil reform on India.[29]

A far more accomplished diplomatist was Spencer Compton Cavendish, Duke of Devonshire, who whilst serving as secretary of state for India took the courtesy title of Marquis of Hartington. After Disraeli's defeat, Hartington declined an invitation to form a Government when Gladstone made it clear he would not serve under any of his colleagues. Hartington in London and Griffon working at closer range, more than anyone else, were the two officials to be credited with successfully negotiating Abdur Rahman's accession to power on British terms. Placing him on the throne, without bloodshed and with tribal consent, marked a major victory not only for Britain, as it was also a piece of good fortune for Afghanistan for the next 21 years.

Lord Ripon arrived at Simla on 8 June 1880 and on 3 July Lord Lytton set sail from Bombay, having left on his desk a lengthy minute for his successor, full of the bitterness of a man whose policy was in ruins. Lytton expressed his vehement opposition to any reconciliation with Yakub Khan for 'his participation in the massacre of the British Embassy'.[30] If Ripon made any attempt to reinstate the former emir, Lytton warned he would 'omit no means or opportunities available to me of opposing and publicly condemning any such action'.[31] What Lytton chose to ignore was the fact that Yakub had been packed off to a cosy exile in India rather than being hauled before Roberts's military tribunal, precisely because the Government lacked any hard evidence of his complicity in Cavagnari's murder.

On 22 July, a great durbar with attendant fanfare was held at the Sherpur cantonment. A multitude of tribal maliks and elders assembled to receive Lepel Griffin's solemn announcement that the Government had reached satisfactory arrangements with Abdur Rahman and, the best news of all, the troops would very shortly be evacuating the country. A letter from Lord Ripon was read to the gathering, spelling out the details of the pact: there would be no British Resident at Kabul (though by mutual accord a Muslim agent would reside at the capital), Abdur Rahman was not to engage in political relations with any foreign power other than the British Government, Britain promised to aid the emir in repelling any foreign aggressor, and, as a gesture of goodwill, the new ruler would receive a grant of up to 1 million rupees and some guns for his defence.

No sooner had the tribesmen began dispersing to celebrate the emir's enthronement than orders were issued for the army to make ready for the march to India. Roberts was to return along the Kurram route, but before leaving he decided to pay a visit to Jalalabad, wishing to inspect the place where Sir Robert Sale, a soldier he much admired, had in part redeemed the failures of the last war. Roberts's intention was to carry on to the Khyber Pass for a look at the scene of Chamberlain's humiliation, the event that had touched off the current war. 'But suddenly a presentiment, which I have never been able to explain to myself, made me retrace my steps and hurry back towards Kabul.'[32] Roberts's foreboding was fully justified, for when he came across General Stewart riding out to meet him, he was given the astonishing news that General George Burrows, who had set out from Kandahar with a brigade to oppose the advance of Ayub Khan from Herat, had suffered a calamitous defeat on 27 July at Maiwand, a small village a few miles east of the Helmand River.

The Battle of Maiwand shocked Victorian England to its foundations, with the memory of the South African disaster at Isandlwana the previous year still fresh in people's memory. The loss of nearly 1,000 British and Indian troops in a single day, at a moment when the public was led to believe that the Afghan war was ended, made a tremendous impact on British society and even found its way into the popular fiction of the day. 'The campaign brought honours and promotion to many, but for me it had nothing but misfortune and disaster. I was removed from my brigade and attached to the Berkshires, with whom I served at the fatal battle of Maiwand.'[33] So opens the account of Dr John H. Watson, late of the Army Medical Department, in Arthur Conan Doyle's first Sherlock Holmes story, *A Study in Scarlet*. It was published eight years after Maiwand, where the good doctor said he took an Afghan bullet in the shoulder.

Not everyone was pleased with Abdur Rahman's accession to the throne. Ayub Khan had been ensconced in Herat since the start of hostilities with the British. He received material assistance from the Persians, who never gave up hope of annexing Herat and whose dislike of Britain stretched back to the shah's humiliation in 1839. This aid enabled Ayub Khan to raise a large lashkar for the march on Kandahar. He started on 15 June with 7,500 men and as the force tramped south across hills and baking desert, the tribesmen poured out of their homes in the thousands to flock to Ayub Khan's banner. By the time he met Burrows's brigade, Ayub Khan's lashkar had swollen to more than 20,000. Burrows had left Kandahar on 4 July with a column of roughly 4,500 fighting men. The force was reduced to just under 2,500 when on 13 July his Afghan contingent deserted to the enemy. Ayub Khan's determination to take Kandahar as a first stage to driving the British from Kabul was no secret. Rumours of an impending attack had been circulating for more than six months. An Afghan nobleman, Sher Ali, had been appointed Wali, or Governor, of Kandahar after the signing of the Treaty of Gandamak. It was this chieftain who brought reports of a call for jihad in Herat and urged the British to send a force to confront Ayub Khan.

Putting aside their traditional rivalries, the various chiefs at Herat were sufficiently incensed at the foreign occupation of Kabul to put up a united front against the infidel. The British possessed no hard intelligence of Ayub Khan's movements until 27 June, by which time he had advanced halfway to the Helmand River. It was now too late to mobilize reinforcements from the Reserve Division on the Indus, so Major General James Primrose, whom Stewart had placed in charge at Kandahar, ordered General Burrows to take a brigade to meet the approaching Afghan army. Burrows reached the Helmand on 7 July, three days after starting out from Kandahar. Sher Ali's troops were deployed on the west bank, with six guns. Two days later levies could no longer resist the mullahs' cry for jihad. They deserted to Ayub Khan and the governor, Sher Ali, rapidly made his way back across the Helmand to become a fugitive in the British camp.

Burrows was able to make a quick crossing of the river, which on account of a drought was fordable everywhere, to retrieve the artillery the Afghans had left behind and carry the guns back to his side of the river. Burrows was not meant to cross the Helmand, staff headquarters believing this would have left the brigade exposed and out of supply range. The river's low water level also made it easy for Ayub Khan to cut off Burrows's retreat. Thus was lost the only chance of success in the forthcoming engagement. Had Burrows taken his men across the river and marched ahead to encounter Ayub Khan's army, 'the clansmen, seeing his forward march, would have remained quiescent and awaited the issue of the battle'.[34] As it was, watching the British fall back from their position on the Helmand emboldened the tribesmen and Ayub Khan was joined by many more fighters before the battle began, virtually sealing Burrows's fate.

On 26 July, Burrows was told that Ayub Khan intended moving on the village of Maiwand, about 80 miles west of Kandahar across open desert, an easy marching terrain that could take him to the town without much difficulty. This would have left the Afghan chief in a position to carry on north up the main route to Ghazni and Kabul, and the result of this would have been a major new phase in the war which the Government had taken as concluded. The brigade broke camp on the morning of 26 July to move on Maiwand, slowed by the need to allow the camels with all their stores to keep pace with the troops.[35] A few hours into the march, Burrows's scouts reported large bodies of horsemen moving towards Maiwand from the north, but the haze and dust of the lashkar prevented any accurate calculation of the enemy's strength. 'It was evident that a collision with Ayub Khan must take place before we reached our destination,' Burrows later put on record. 'It was difficult . . . to estimate the numbers of the enemy, but judging from the extent of the country covered, I believe I am well within the mark when I set down the strength of twenty-five thousand.'[36] That gave Ayub Khan more than ten times the strength of the British brigade. Burrows was under heavy pressure from the commander-in-chief, General Sir Frederick Haines, to adopt all possible means to prevent Ayub Khan reaching Ghazni. The strategy Burrows decided on was to intercept the enemy at the village of Maiwand.

The brigade went into action just before noon on 27 July, in less than ideal conditions. Burrows himself had never commanded in battle and he had no experience of a mixed force of infantry, cavalry and artillery. A third of his infantry strength was the 30th Regiment of Bombay Native Infantry, known as Jacob's Rifles, a unit comprised of recruits with little or no training. The battle was fought on open ground, where the British should have given a good account of themselves, instead of in the mountain country that is the Afghans' natural fighting habitat. But the heat was excessive for men carrying full battle kit and heavy ammunition pouches. The troops had been roused at midnight and worked for six hours to prepare themselves and the animals for the 12-mile march. Many started off on empty stomachs and the supply of water failed early in the battle.

The brigade rapidly moved on Maiwand, when through the haze a large body of men was spotted advancing diagonally across the right front. Burrows shifted his troops to higher ground, where he deployed his infantry in the centre, the cavalry on the left, covered by two guns and a reserve troop of cavalry. Many of the men were already half-mad with thirst, apart from being hopelessly outnumbered and outgunned by more than 30 Afghan artillery pieces. Ayub Khan's cavalry was now observed through binoculars trotting on at a distance of less than three miles. Burrows's spies reported that the Afghans' main body had already occupied Maiwand and were moving into position to cut off any possible British retreat. The general determined no time was to be lost – he gave the order to attack at once, the 'advance' was sounded and the brigade moved forward. As the British advanced the Afghan cavalry fell back in good order, and nothing was as yet seen of the force that held Maiwand.

Lieutenant Hector Maclaine opened the three-hour battle by advancing his cavalry across a *nullah*[37] and charging the enemy at a gallop, a manoeuvre that set the scene for a day of costly mishaps. Maclaine shifted his guns closer to the front than had been intended. He had to wait for the infantry and cavalry to come up in support in a line, with the 66th of Foot on the right, Jacob's Rifles taking the centre and the 1st Grenadiers on the left. Maclaine's tactical errors cost him dearly: when he went off to search for water for the wounded men he was taken prisoner by the Afghans and was not seen again. A burst of explosions rocked the plain, signalling the commencement of a heavy artillery duel that carried on for more than an hour. The British were hopelessly outgunned: the Afghan possessed better and heavier artillery, including six state-of-the-art Armstrong guns, a British rifling breech-loading field gun. The Grenadiers on the left flank were unable to find cover and were being blown to pieces by the enemy's more powerful guns. A chill went down the men's spines when the Afghan infantry appeared in strength, numbering in the thousands.

The enemy stood in front of the British line, massing for an assault. Burrows sent the Grenadiers forward but then cancelled the order when he realized how severely they had been mauled in the artillery barrage. In the confusion of battle, Jacob's Rifles rushed across open ground with two smooth bore guns to halt the enveloping Afghan cavalry. What Burrows and his aides didn't know was that a second *nullah* ran beside his other flank, joining the main ravine in his right rear. A horde of Afghans raced along this *nullah* to infiltrate the British right flank, forcing the 66th of Foot to turn at right angles to Jacob's Rifles and the Grenadiers. This left Burrows's force strung out in a horseshoe formation, with Afghan cavalry on the left flank and the enemy's guns pounding the brigade on the right.

The two companies of Jacob's Rifles came in for an unpleasant time when the smooth bore guns ran out of ammunition, with no chance of bringing up more from the rear. The men were left to swelter in the overpowering heat and withstand the artillery bombardment on open ground. With the British guns out of action, the Afghan cavalry moved in behind their left flank, whilst their infantry continued to slip unmolested along the *nullah* on the right. The enemy guns opened on the Grenadiers at a devastating 300 yards' distance. Ayub Khan next threw several thousand tribesmen against the British infantry line, with deadly effect. Jacob's Rifles broke and fled the field on the left, leaving the Grenadiers to take the brunt of the attack. They were cut to pieces, many of the men simply too exhausted to put up a fight. The Royal Horse Artillery guns positioned in the centre were hastily withdrawn, at which point the remainder of Jacob's Rifles dissolved into the 66th of Foot's line, throwing the right flank into disarray. Burrows rode about the field trying to raise the troops to a cavalry attack, but they were not up to it. The infantry broke to the rear in different directions. Many of the Grenadiers were killed in the retreat and the 66th fell back, fighting in small groups. The remnants of the Grenadiers and the 66th clustered together for a stand on the edge of the nearby village of Khig. The survivors were forced out of the village and grouped for a second stand in a walled enclosure, where all but 11 of the original 100 defenders remained alive.

The final stand was made by two officers and nine soldiers of the 66th outside the village. The regiment's terrier Bobby was wounded in the massacre but escaped to join the retreat to

Kandahar. From there the dog was sent back to England, where Queen Victoria at Osborne House presented her and other survivors of the battle with the Afghan War campaign medal.

What was left of the brigade and camp followers disbanded in a wild retreat up the road to Kandahar, the Afghan cavalry in hot pursuit. Burrows made his escape after gallantly giving up his horse to a wounded officer. The survivors struggled on more than 40 miles to Kandahar. When around 3 a.m. on 28 July Primrose learnt that the brigade and many stragglers were fighting their way towards Kandahar, he ordered a relief force of some 500 to give a hand to the beaten army. The Afghans, too fatigued by fighting to confront a fresh British column, wheeled their horses round and withdrew. They had no way of knowing that the garrison held only 1,800 men. To his credit Burrows, who that day had two horses shot under him in the engagement, was amongst the last to reach Kandahar. The Battle of Maiwand claimed 1,109 British and Indian casualties, according to the most reliable estimates, of which 969 were killed, a third of the brigade's total strength. Afghan losses amounted to at least 3,000, and possibly more.

The Maiwand disaster drew a thunderous overreaction from Government, with Ripon threatening action against those responsible for what he condemned as 'meagre and unsatisfactory' dispatches that left the Government in ignorance of the facts of the battle and 'the real reason of the reverse'. Haines, the commander-in-chief, slated his subordinates in the field for failing to ascertain the true strength and position of Ayub Khan. Haines believed that victory had been possible 'had the infantry stood firm, and that the disaster was intensified by the demoralisation of the cavalry and the precipitous retreat'.[38] The press, in India and at home, jumped on the general condemnation of those in charge at Kandahar and Maiwand:

> *The Bombay Gazette* states, on the authority of private letters, that General Burrows' troops were so demoralised by the heavy artillery fire ... that they could not be got to stand the rush of

the Ghazis. The *Calcutta Englishman* adds, on similar authorities, that the 66th Foot were almost as bad. It would be well to await fuller information before we lay blame on the troops, and when we remember that General Burrows had never before seen a shot fired in anger, and that General Primrose has shown no evidence of military capacity but rather the reverse, it would seem that the responsibility must eventually fall in a higher quarter than on the private soldiers.[39]

Should Burrows and Primrose, who failed to obtain vital intelligence to better frame their strategy and hence save men's lives, be held accountable for the catastrophe at Maiwand? What of the Government leaders, so desperate to prevent Ayub Khan moving on Kandahar and Kabul, an event that would have brought fatal political consequences in Calcutta as well as London? Or taking the saga back to its origins, should the blame for the entire debacle be laid at the feet of those who sanctioned the invasion of Afghanistan? The inescapable truth is that for the second time in less than 40 years, a lethal cocktail of incompetence and arrogance at official level had brought ruin on a British army. Once again, the hour of retribution was at hand.

As we have seen, Roberts was riding back to Kabul from an abortive trip to the Khyber when he was met by General Stewart bringing up news of the disaster. 'This lamentable story imparted to me by Stewart almost took my breath away, and we eagerly discussed the situation as we rode back to Sherpur.'[40] The two generals agreed that the only means of affording speedy relief to the Kandahar garrison was by sending a large force from Kabul.

Ayub Khan had broken off his pursuit of the fleeing army, but he vowed to return and finish the job at Kandahar, and he was as good as his word. On 29 July the garrison began the work of strengthening the citadel's defences, cutting down walls and building enclosures – a job, it might be argued, which should have been undertaken by Stewart when he commanded at Kandahar. Following Nott's example in the previous war, Primrose had a great number of Afghans and their families turned out of the city to facilitate military operations and reduce the number of mouths to feed. A telegraph line was laid round the walls connecting all the principal posts. The garrison dug in for the siege that all expected, and they didn't have long to wait.

On 5 August, advanced guards of Ayub Khan's army showed themselves outside the walls, accompanied by a throng of fanatical ghazis. The Afghans took possession of the surrounding hills and cantonments that had been abandoned when Primrose ordered his force to withdraw behind the walls. Heavy fire from the Afghan positions and occasional sorties to dislodge the enemy from nearby villages and fortifications marked the tenor of the next days. The troops not only defended themselves with valour but more than once went on the offensive, with admirable results. It did not take long for Ayub Khan's ghazis to begin to waver and melt away. Some of the Afghan regular troops were in a mutinous mood over the lack of success and salary arrears. Hence almost no effort was made to press the siege, which practically came to an end after a fortnight, leaving the Afghans little hope of capturing Kandahar. On 22 August, Ayub Khan shifted strategy when he was told of a British force marching from Kabul. He left his position near Kandahar to encamp behind the village of Pir Paimal, some three miles west of the city.

Lord Ripon and his cabinet were in complete agreement on the need for immediate action. The question was who should be sent to Kandahar. The Government at first thought of the Quetta troops, an idea strongly opposed by Roberts, who was champing at the bit. His reasoning, apart from a strong wish to lead the force himself, was that compared with the battle-hardened Kabul garrison, the less professional Bombay Native Infantry was not up to the task. Moreover, it would take 15 days at least to equip the troops at Quetta, whereas the Kabul force was ready to move at once. The General detected an uneasiness taking hold in India, where the Maiwand disaster was bringing to the surface a mood similar to that of the pre-Mutiny days. Roberts's view prevailed in the end. With Ripon's consent, Stewart gave the command to Little Bobs, who was to march south with a column of 10,000 men. By 8 August he was ready to move out. Roberts was aware that the eyes of India as well as England were upon him, though it is doubtful he could have envisaged the honours in store for him. 'Kabul to Kandahar' became a byword for gallantry and captured the public's imagination as nothing else had done in this war. Fortunately for Roberts the Kandahar garrison was still occupying Ayub khan Khan's attention, thus clearing the way for an unopposed, history-making march of 313 miles in 23 days.

Two days into the expedition, on 11 August, Roberts received his last communication in the field. It was a telegram from his wife, Lady Nora Roberts, from a little village in Somerset, wishing the force God's speed. From there Roberts struck out onto the barren plain, having to contend with 110°F heat, 940 men too sick or lame to walk, the scarcity of provisions and forage for this vast force, and he himself in pain and weak with fever. But on 30 August he was able to telegraph Simla that he had arrived at Mohmand, a village barely a day's stage from Kandahar. The next morning the Kandahar commanders came out to greet the column and Roberts, riding into the city, took note of the 4,000-strong garrison, the 15 guns, the 30-foot-high walls, and wondered what all the fuss was about. 'For British soldiers to have contemplated the possibility of Kandahar being taken by an Afghan army showed what a miserable state of depression and demoralisation they were in.'[41]

Having rested his men at Mohmand on the penultimate day of the march, Roberts lost no time in reconnoitring the terrain on which he would have to meet Ayub Khan. The information the general gathered on his reconnaissance, which was not without a brief but intense skirmish with the Afghan troops, decided him to engage the enemy without delay the following day. The evening of 31 August, two brigades moved off to occupy the hills nearby Kandahar, whilst Roberts took up a position to the west of the city. This placed him within striking distance of Ayub Khan's camp. On the morning of 1 September, Robert formally assumed command of the army in southern Afghanistan, the total force numbering nearly 15,000 men with 36 guns. The battle was a short, sharp engagement lasting a few hours. It ended by late afternoon, with a stunning victory for Roberts, the complete rout of the Afghans and the capture of all their guns.

Soon after 9 a.m. the 40-pounders opened up on the main Afghan camp. This was followed by a charge of the 92nd at bayonet point to clear a key village stronghold. The 72nd and 2nd Sikhs, supported by the 5th Gurkhas, took another strategic village and the brigades then joined for an assault on the Pir Paimal ridge. This was rushed by the Highlanders and Gurkhas, who dispersed a mass of 8,000 Afghans. The camp was entered at 1 p.m. and found to be deserted. The tents were left standing and everything had been hastily abandoned, from meat in the cooking pots to half-kneaded bread. The Afghans had found time to murder their prisoner Lieutenant Maclaine, who was discovered at the camp with his throat cut. As usual, it was difficult to estimate the number of enemy killed, but the British buried 600 bodies between Kandahar and Pir Paimal alone. British losses on the day numbered only 40 killed.

When the British departed Afghanistan later that month, Ayub Khan made another, and this time almost successful attempt to seize Kandahar. But he was confronted in person by the new emir, Abdur Rahman, who took the field and delivered Ayub Khan a crushing defeat. He fled to Persia and in 1888, in the grand British tradition of offering asylum to their former enemies, Ayub Khan became a pensioner in India.

The Second Afghan War was over. The last British troops left Afghanistan in a phased withdrawal on 21 April 1881. British India had come out of this conflict, like its predecessor, with heavy loss of life, calculated at 50,000 casualties, a gaping hole in the Government's finances, large swathes of land laid waste and its inhabitants left impoverished, and with the enduring enmity of the Afghan people. Britain retained the strategic Pishin and Sibi valleys in Baluchistan. The Khyber Pass and the Kurram Valley, through which ran the two main approaches to Kabul from India, were placed under British administration as arranged by Lytton in the Treaty of Gandamak. In reality, they reverted to tribal control and later became the scenes of fierce fighting with British troops.

If anyone emerged victorious in this war it was Little Bobs, the hero of Kandahar. In 1893 he left India for the last time, after 41 years' service under the Raj. He was lionized by his countrymen on his return, he was the Queen's guest at Windsor, Eton and the City of London presented him with swords of honour, and he became the inspiration for a special medal, the Roberts Star, awarded to all who had marched with him to Kandahar, including his horse. Roberts was raised to the peerage as Baron Roberts of Kandahar. He was later made a field marshal and appointed commander-in-chief in Ireland. At nearly 70 he was once more sent out on active service, this time to the Transvaal, where he directed operations against the Boers. In the First World War, Roberts sailed to France in November 1914, at the age of 82, to be with the men of the Indian Army Corps. He caught a chill during the Channel crossing and died on landing at St Omer.

CHAPTER 8

Once More unto the Breach

Three events took place during Abdur Rahman's reign that drew Britain back into Afghanistan's gravitational orbit. The first incident, in 1885, came as a veritable gift to the Forward Policy advocates. The Panjdeh affair brought Britain closest to war with Russia than it had ever been in all the years of Great Gamesmanship. Encouraged by the complete withdrawal of British troops from Afghanistan in 1881, three years later the tsar's army occupied Merv, only 130 miles from the vaguely defined Afghan border. This bold advance conceived by General Mikhail Skobolev took the two imperial powers even closer to the brink than at the time of the Russian-backed siege of Herat in 1839. The flashpoint was the remote oasis of Panjdeh, lying halfway between Merv and Herat, and only 12 miles from the Afghan outposts. When it was discovered that Russian forces were threatening Pandjeh, Abdur Rahman rushed a contingent of troops north to defend the oasis. Then in March 1885, the British ambassador in St Petersburg, Sir Edward Thornton, delivered a note to the Russian Foreign Minister Nicolay Girs, cautioning that a Russian attack on Panjdeh might lead to the most disastrous of consequences. Thorton later turned up the heat when he informed Girs, in unadorned language, that British prime minister William Gladstone would meet any attempt by Russia to approach or occupy Herat, which is not far from Panjdeh, with a declaration of war.

Russia took a gamble by disregarding the British warning and on 30 March, the tsar's troops attacked and destroyed the Afghan garrison at Panjdeh. 'All through April, the danger of war was imminent, not so much on account of the Panjdeh affair itself, as because it seemed to foreshadow in the popular view a further Russian advance.'[1] The Government of India mobilized two army corps for deployment to Afghanistan, in compliance with their treaty obligations with that country, and ultimately as an act of self-defence. But the Russians, as history has shown, are masters at pushing their adversaries to the limit before initiating a tactical withdrawal. This was seen at Herat in 1839, as it was 123 years later in the Cuban Missile Crisis. In 1885, the Russians halted their advance as soon as they had secured a compromise deal with Britain which acknowledged de facto the tsar's territorial gains up to that point. The agreement took the form of a protocol signed in 1887, demarcating the Russo-Afghan frontier, roughly along the lines agreed in a treaty signed 34 years earlier in which Russia recognized Badakhshan and Wakhan as part of the domains of the Afghan emir. This stopped the Russian advance dead in its tracks for almost a century, until 1979 when Red Army tanks rolled across the Oxus into Afghanistan.

The Panjdeh crisis drove home the need for firmly established boundaries between British India and Afghanistan. This task fell to the career diplomat and the Government of India's foreign secretary, Sir Mortimer Durand, who happened to be the son of Henry Durand, the soldier who blew open the gates of Ghazni in the First Afghan War. The strategic question of Afghanistan's western frontier with Persia was settled with the signing of the 1887 treaty. There remained the southern and eastern limits of the emir's dominions. Durand maintained a close personal relationship with Abdur Rahman and in 1885 he personally informed him of the massacre of the Afghan garrison at Panjdeh, a piece of intelligence the emir dismissed with a shrug.

The Durand Line, as the border came to be known, is an illogical, contentious boundary stretching for more than 1,500 miles and that was negotiated in a matter of weeks. 'It cuts across one of the main basins of the Indus watershed, it splits a nation in two, and it even divides tribes.'[2] The Durand Line has been exploited by every Afghan ruler since Abdur Rahman to whip up public resentment against British India, and afterwards Pakistan. Yet the Durand Line gave Britain a point of reference for defining what constitutes a border violation, such as Afghanistan's 1919 foray into the Khyber Pass.

The other signal event that took place between the Second and Third Afghan Wars involved Afghanistan as an indirect participant. This was the great Pathan Revolt of 1897, a tribal uprising that turned into a serious challenge to British rule on the Frontier. In the summer of 1896, a gun battle erupted in a village in the parched badlands of North Waziristan when the British political officer arrived to mediate in a local dispute. The shootout cost the lives of 13 Indian soldiers, and to punish this outrage, two brigades of native levies advanced into the region, destroying villages as they went, in application of the policy of collective responsibility. The intractable Wazirs refused to submit, and by the following year the flames of rebellion had spread like wildfire up and down the Frontier. This was the year of Queen Victoria's Diamond Jubilee, a time when the British public was indulging in an outpouring of imperial fervour. All the more shocking then was the news that British India, the Jewel in the Crown, had been placed on a war footing in a remote corner of the Empire which everyone took for granted was locked securely under British control.

The Malakand garrison in the Frontier's northernmost reaches came under attack on 26 July 1897 by 10,000 tribesmen under the banner of Saidullah, known to the British as the 'Mad Mullah'. The young journalist Winston Churchill was present at the siege, which he described in lively detail:

> In the attack on the Malakand camp all the elements of danger and disorder were displayed. The surprise, the darkness, the confused and broken nature of the ground, the unknown numbers of the enemy, their merciless ferocity, every appalling circumstance was present. But there were men who were equal to the occasion.[3]

Foremost amongst these men was the superbly named General Sir Bindon Blood, commander of the Malakand Field Force, who arrived just in time to rescue the garrison from certain annihilation. It is significant to note that the scheming hand of Afghanistan was never far from the scene throughout the tribal uprising. Abdur Rahman was angry over the Government of India's stated intention to bring the Frontier hill tribes closer under British dominion, and he was equally unhappy with the way the Durand Line treaty had, in his view, been forced upon him. A gathering of mullahs was convened in Kabul ahead of the uprising to call for the expulsion of the Feringhees from the Frontier. Mullahs like Saidullah were dispatched on a holy mission to raise the tribes to rebellion; money and arms were distributed to the tribesmen; and all this with the connivance of the emir.

Some of the actions to suppress the revolt were taken at division level, with General Sir William Lockhart's Tirah Field Force alone comprised of 34,500 troops, 20,000 camp followers and 72,000 baggage animals, a force larger than the Army of the Indus that had marched into Afghanistan in 1839. A host of immortal regiments were thrown into this campaign, from the Gordon Highlanders and the Bengal Lancers, to Hodson's Horse and the Gurkhas. It was an army worthy of Empire and after nearly two years of fighting, the Frontier was pacified, at a cost of 1,150 British and Indian casualties. For a few months, there was a real risk of Peshawar being overrun and it looked like the British might have suffered the ignominy of being pushed back to the Indus. The lashkars counted their fighters in the tens of thousands, but in the end the unruly war parties were broken up by Lockhart's disciplined, better-equipped troops. Peace had been restored, though 'truce' would be a more accurate description of the situation that prevailed at the conclusion of the Pathan Revolt. On the North-West Frontier, periods of peace could never be taken as anything but a lull between hostilities.

By Western criteria, Abdur Rahman had acted in a highly duplicitous fashion by encouraging the religious zealots to whip up the Frontier tribes against the British. This was in flagrant violation of the spirit as well as the letter of the Treaty of Gandamak and the Durand Line protocol. Taking it back to 1855, the emir's grandfather Dost Mohammed had signed a treaty in which he pledged to be 'the friend of the friends and enemy of the enemies of the Honourable East India Company'. The Emir of Afghanistan, like his predecessors and those who came after, could subordinate his contractual obligations to political expediencies effortlessly and without suffering the slightest pangs of conscience. Britain's usefulness was chiefly as a powerful military force, a deterrent to a Russian invasion of Afghanistan. Once that threat had ceased to exist, in the emir's perception, treaties became a secondary consideration.

Abdur Rahman was an energetic, clever ruler who treated his loyal subjects with kindness and his opponents with summary execution. Lepel Griffin described him as a man with an exceedingly intelligent face and frank and courteous manners, shrewd and competent in conversation and able to get on with the business in hand. Abdur Rahman also rivalled the great medieval Central Asian conquerors in his lust for cruelty. During his reign, crime was kept within bounds as a result of ruthless punishments, mutilations, blindings and the death penalty, all of which were frequently inflicted in the emir's presence in open durbar. The most ghastly penalty was imprisonment in the Black Well, into which condemned prisoners were flung and left to die amongst decaying corpses, filth, reptiles and vermin, food and water being lowered at intervals to prolong the agony of the captives. Abdur Rahman's treatment of criminals and dissenters would have struck terror into the hearts of the Taliban. Yet along with Ahmad Shah and Dost Mohammed, the emir that Britain had placed on the throne of Kabul before quitting Afghanistan achieved the status of national hero.

On the positive side, during his 21-year reign, from 1880 to 1901, Abdur Rahman took his country towards, if not quite into, the twentieth century. The 'Iron Emir' made strident efforts to modernize Afghanistan and turn his divided people into a coherent state. During these years, Afghanistan acquired its current borders, with the Durand Line and the Wakhan Corridor, the elongated mountainous valley between the Hindu Kush and the Pamirs, set up as a buffer to prevent the British and Russian Empires meeting head on. Abdur Rahman reined in the power of the mullahs, whom he dismissed as ignorant despots. He also attempted to break the tribal system and bring the maliks under his control. One example of this was a decree in favour of women's rights, and this in nineteenth-century Afghanistan! Under tribal custom, if a woman lost her husband the deceased's next of kin had the right to marry her, even against her wishes. The emir took the bold step of issuing a decree freeing widows to marry whomever they wanted.

Abdur Rahman was a diabolically clever and ruthless man. In order to deal with the rebellious Ghilzais, he transplanted thousands of families north of the Hindu Kush in a forced migration, which had the dual effect of diluting their power in the south whilst enhancing Pashtun influence amongst the Uzbeks and Tajiks of the north. Abdur Rahman's reign was often marked by brutality, but he also achieved significant political reforms, such as organizing his government into departments along Western lines. It can safely be said that he was the first Afghan ruler to bring his country under a centralized administration. The Government of India was delighted to have in power an Afghan emir who not only displayed friendship towards the British, but who could also be trusted to abide by his treaty obligations, in particular to refrain from dealing with any foreign power other than Britain.

One of Abdur Rahman's reforms that would later excite less enthusiasm in British India was the emir's overhaul of the armed forces. The Afghan Army, which in the past was simply a mass of fighting men, was in the late nineteenth century brought up to professional standards, with the institution of rigorous Western-style officer training and exams. Unhappily for the British, it became a far more formidable fighting force.

Abdur Rahman died in 1901 and as a mark of the emir's firm

hold over the country and its institutions, his son Habibullah inherited the throne in a remarkably seamless transition. Habibullah began his reign by attempting to renege on Abdur Rahman's commitments with Britain. 'At first he refused to renew the agreements entered into by the former Emir and commenced intriguing with the Afridis. It was not until 1905 that he fell in with Abdur Rahman's policy.'[4] That was when a mission led by Sir William Louis Dane, the Government of India's foreign secretary, persuaded him to reaffirm his father's engagements, including the controversial Durand Line that defined Afghanistan's eastern border. In exchange, the emir was granted an annual subsidy of 1.8 million rupees, some £400,000. Habibullah pressed on with his father's policies of modernization, with factories, hydroelectric plants and road-building schemes.

The reformist effort proved an uphill battle for the emir, who all along had to struggle with reactionary forces at home and abroad, notably from Turkey and Germany during the First World War. Those who opposed his dangerously secular tendencies accused him of betraying Islam when he refused to call a jihad against British India. One night in February 1919, whilst Habibullah was on a shooting trip in the Laghman Valley near Jalalabad, a gunman crept into the emir's tent and blew off the top of his head with a rifle. The killer's identity remains a mystery to this day, but at the time the finger was pointed at a variety of suspects, from the British to the Russians, as well as a raft of domestic rivals. Habibullah's assassination was in all probability engineered by elements opposed to his reformist work and his 'soft' stance on British India. Popular outrage somehow had to be placated, therefore an Afghan officer, Colonel Shah Ali Raza, was charged with the murder, convicted on the flimsiest of evidence and bayoneted on orders of the court of inquiry. Immediately after the assassination, Habibullah's brother Nasrullah, a confirmed fundamentalist in today's parlance, proclaimed himself emir in defiance of the rival claim to the throne by Amanullah, Habibullah's third son. As governor of Kabul, Amanullah had access to the city's treasury and arsenal and assembled the tribal notables and top military commanders, who proclaimed him emir on 28 February 1919.

The army was incensed over Habibullah's murder and Nasrullah, keen to avoid any suspicion of complicity in the crime, travelled from his Kandahar stronghold to Kabul to pay homage to the new king. For his efforts, Nasrullah was arrested and declared to be guilty of instigating the murder of the late emir. He was sentenced to imprisonment for life, dying shortly afterwards. Amanullah's brothers were likewise confined to prison or placed under house arrest. Amanullah was a 27-year-old jovial character who enjoyed strutting about in top-boots and whipcord riding breeches and was thought by the British to be conceited, arrogant and somewhat dim. Powerful elements within the army harboured suspicions regarding the emir's involvement in his father's death. Amanullah attempted to purchase their loyalty, as well as time for himself, by granting the officer corps a substantial pay rise. But pressure was mounting on the emir to appease his increasingly nationalistic generals, many of whom were demanding action against the British infidels who had the audacity to occupy Muslim land across the border. The anti-British party was led by a group of senior officers and the dowager queen, Amanullah's mother. Having no desire to follow in his father's footsteps, Amanullah came out to declare himself a staunch supporter of the Islamists. The commander-in-chief, Nadir Khan, who was suspected of complicity in Habibullah's murder, was removed to Khost in eastern Afghanistan, with General Saleh Mohammed appointed in his place.

Amanullah, like his father Habibullah, kept friendly relations with British India. Habibullah had resisted pressure from the Central Powers to open hostilities with Britain in the First World War. A German mission, accompanied by representatives of the Sultan of Turkey, turned up in Kabul in 1915 to press on the emir a treaty to provide Britain's enemies with material aid. Habibullah was adept at playing both sides for the best deal. He accepted military assistance from Britain's enemies but politely dismissed their invitation to declare jihad on his neighbour, and the Germans left Afghanistan in disgust. For the moment, this had the effect of silencing the anti-British agitators in the army and amongst the religious leaders – but only temporarily.

On his accession to the throne, Amanullah wrote to the viceroy, Lord Chelmsford, proclaiming his friendship and eagerness to enter into treaties from which both countries might derive commercial benefits. In doing so, Amanullah ironically set the scene for armed confrontation with Britain. 'The action of Amanullah in condemning his uncle, the champion of the mullahs ... alienated alike the mullahs and the army.'[5] Discontent spread amongst these disaffected factions, with the result that the *khutba*, or public prayers, were not read that month in the emir's name, a deliberate affront to his authority. Amanullah was left desperately seeking a means to gain support from his people and his generals. He found it in the traditional expedient of making war on an unpopular neighbour, and in doing so he showed considerable astuteness:

> While on the one hand he was able to pose as the leader of his people in their march to freedom from British suzerainty, on the other he appealed to their religious fervour by proclaiming a holy war against the unbelievers and to their cupidity by holding out to them the fair prospect of loot which an invasion of India would furnish to his followers as it had done to their forefathers from the days of Mahmud of Ghazni.[6]

The British Government was in no position to devote much time or resources to analysing and countering warning signals from across the border. Nor did they have any great awareness that plans were being drawn up to attack British territory. In fact, Major George Roos-Keppel, serving as Chief Commissioner of the North-West Frontier Province, wrote to the viceroy in one of his weekly dispatches in April that he was looking forward to a 'quiet summer'. Roos-Keppel, of all people, a soldier who had commanded the Khyber Rifles for 12 years and who conversed with the Pashtun tribesmen fluently in their own language, should have picked up the signs of unrest.

At this time, the British almost went out of their way to provide Amanullah with ammunition to call for a holy war. The tribal durbar convened to try the unfortunate Colonel Raza and Amanullah's uncle Nasrullah coincided with the massacre at Amritsar. This lamentable event took place on 13 April 1919, when Mahatma Gandhi had just stepped into the leadership of the Indian nationalist movement. The 50-year-old London-barrister-turned-spiritual leader organized a series of protests against the Rowlatt Act, emergency legislation that extended indefinitely a set of anti-sedition measures that were enacted in the First World War. The protests turned into riots in many places, including the city of Amritsar in the Punjab, when Brigadier Reginald Dyer ordered his troops to turn their guns on an unlawful assembly of 20,000 demonstrators, officially killing 379 and wounding more than a 1,200 others.[7] Add to this Britain's weakened military and economic condition less than six months after the end of hostilities in Europe, Afghanistan's perennial indignation over Feringhees lording it over Pashtun tribesmen on the Indian Frontier, the clamour for the reinstatement of Afghan control of the country's foreign policy and the alleged injustice of the Durand Line having been 'forcibly' imposed on Abdur Rahman – and Amanullah suddenly found himself incapable of resisting the war party's demands for jihad.

The emir and his supporters began to rally support for the forthcoming struggle and attempted to make use of nationalist sentiment in India and the border tribes' traditional resentment of the British presence in their homeland. 'They campaigned vigorously amongst the Pathan tribesmen, spreading nationalist propaganda and distributing arms in the tribal belt. These efforts were greeted with great enthusiasm by the Afghan public and raised the hopes of the army and of those border tribes that favoured total Afghan independence.'[8] Afghan leaders urged those who considered themselves true Muslims and patriots to prepare for the struggle with Britain, condemned as the traditional foe of Afghan independence.

This was the moment for Amanullah to establish his Islamist credentials with a highly suspicious military and clerical hierarchy, many of whom were not convinced of the emir's innocence in his father's murder. A wave of anti-colonialist agitation was sweeping India and the economy had taken a severe battering in the Great War, especially for the poorer classes who found themselves struggling with a surge in food prices and shortages of many basic goods. 'On the political front, wartime sacrifices had increased the pressure for early Home Rule', and at Lahore, the Congress Party and the Muslim League had joined forces to press their demands for self-government. 'Finally, the war had shattered belief in British invincibility.'[9]

In early May 1919 the Peshawar postmaster, an Afghan named Ghulam Haidar, was in Kabul briefing the emir on the prospects for an uprising against the British. Afghanistan had never relinquished its claim on Peshawar and Amanullah received with joy reports of instability in the streets and weakness on the part of the Government in dealing with these disturbances. This might be his golden opportunity to succeed where all his powerful predecessors, starting with Dost Mohammed, had met with failure. The emir was encouraged by what he learnt from this civil servant: that the British were militarily poorly equipped to confront a popular revolt (this was true) and that thousands of Peshawar's residents stood ready to raise a lashkar of liberation (this was not true). Amanullah happily agreed to provide the postmaster with bundles of pamphlets calling upon the enemies of British imperialism to join the struggle of their Afghan comrades, who would soon be coming to liberate them. But on his return to Peshawar, Haidar was imprisoned and the pamphlets were confiscated. The postmaster had been betrayed by one of his own staff, who reported the plot to Roos-Keppel.

The Khyber Rifles were called in from their garrison at Landi Kotal to throw a cordon around Peshawar. The authorities threatened to cut off the city's water supply if the conspirators so much as moved a finger. This was an unwelcome prospect, as the month of May is one of the hottest and driest of the year. This policy of imposing collective responsibility met with some success during the years of British rule on the Frontier. It became routine procedure for the Government to threaten to block supplies to a village or prevent its men from entering the Settled Districts to trade if the tribesmen failed to deliver criminal elements responsible for committing outrages. The tactic worked that day in Peshawar, when the plotters were denounced and handed over to the police. Roos-Keppel had acted resolutely, since by this time it had become apparent that the Government had a very serious problem on its hands. Trouble was on its way even before the postmaster had returned to Peshawar.

Amanullah had made intensive preparations for the forthcoming conflict, though it is highly questionable that he ever intended it to develop beyond an exercise in border harassment. The emir campaigned vigorously for jihad amongst the Pashtun tribesmen of the Frontier, spreading nationalist propaganda and distributing arms in the tribal belt. Like his father, Amanullah declared that Afghanistan should no longer be bound by the Treaty of Gandamak, under which Britain was left in charge of the country's foreign policy. Amanullah had planned to send his troops across the border on 8 May, to coincide with the anticipated uprising in Peshawar. He might have received intelligence of the postmaster's arrest, though there is no documentary evidence to substantiate this. It is more likely that his generals either had no knowledge of the postmaster's mission or had decided to act independently, ignoring the emir's authority, for the date of the invasion was brought forward to 3 May.

By the time the postmaster was back in Peshawar, Afghan army units had slipped across the border and swiftly stormed and occupied the village of Bagh at the western end of the Khyber Pass. With the advantage of surprise on their side, the Afghans took their first objective virtually unopposed. The advance of Afghan regulars towards the border had been in evidence for days before the Peshawar postmaster affair. General Saleh Mohammed led an escort of two companies of infantry and two guns to Dakka, a town almost touching the frontier, but the Government of India chose not to respond to this provocation. Saleh then sent his troops across the Durand Line, where they took up positions at Bagh. Two thousand troops followed the commander-in-chief from Kabul to Dakka, 1,500 were deployed to Kandahar, and another 2,000 were marched to Khost, on the border 93 miles south of Kabul. Bagh had a strategic importance as it was the site of the pumping station for Landi Kotal at the top of the Khyber Pass where the Khyber Rifles, the first line of Frontier defence, were garrisoned. With the pumping station out of commission due to the murder of its staff, the garrison was entirely dependent for its water supply on two reserve tanks. Bagh itself was tenuously protected by only two companies of sepoys.

By this time the Government was fully aware of Amanullah's intentions. On 6 May the viceroy telegraphed Secretary of State for India Edwin Samuel Montagu, to the effect that Amanullah had called a special durbar at which, 'weeping bitterly', he read out seditious letters from Indian nationalists. Chelmsford reported the emir as saying, 'I ask you: are you prepared for Holy War? If so, gird up your loins, the time has come.' Chelmsford then refers to the operation at Bagh as 'a clear challenge to us, and the tribes so regard it'.[10] The Khyber Rifles' garrison had not yet come under attack, but the levies would be unable to hold out for long without urgent reinforcements. It was also questionable how long their loyalty would hold out, once the Afghans sent the mullahs to stir up religious fervour against the Feringhees.

Roos-Keppel was not eager to wage war on the Pashtuns, even those of a hostile army from the other side of the border. As a career soldier on the Frontier he had fought the Pashtuns in several engagements, but painfully, for Roos-Keppel, who answered to the model of the soldier-scholar, was a passionate admirer of the Pashtuns.[11] He had written a Pashtun language grammar, he translated Pashtun poetry and was co-founder of Islamia College, today the undergraduate school of Peshawar University. He epitomized those servants of the Raj, administrators and military alike, who came under the Frontier's romantic spell. A strong-minded and charismatic man, he developed a detailed knowledge of tribal society and customs and various Pashtu dialects. There is no doubt that Roos-Keppel, along with Colonel Sir Robert Warburton, played a vital role in keeping order amongst the Pashtun hill tribes of the Frontier.

Roos-Keppel realized that the Bagh occupation was of greater consequence than a border incident. 'While the full extent of Afghan intentions was not known, Roos-Keppel nevertheless advised [Viceroy] Chelmsford to eject the Afghan troops from Bagh before local tribes rose in their support.'[12] If proof of the seriousness of the situation were needed, in less than a week the Afghans had cut the water supply to Landi Kotal; a Khyber Rifles patrol escorting a camel caravan up the pass had been halted by armed Afghan pickets who threatened to fire if the party advanced; a group of roadside labourers was murdered in cold blood; and Afghans of the Mohamand and Shinwari had poured into Jalalabad to be given rifles with which to wage jihad. On 5 May regular Indian Army troops under Lieutenant Colonel John Willans were rushed from the Thal garrison in the Punjab to Parachinar, the capital of Kurram. The next day, 6 May 1919, general mobilization was ordered and British India declared war on Afghanistan.

The military position in the Third Afghan War placed British India at a disadvantage in terms of the number of trained units available for immediate service on the North-West Frontier. The Afghan Army was not a very formidable force, though since the reforms undertaken by Abdur Rahman in the late nineteenth century it had been fashioned into a far more disciplined and better-equipped adversary than the ragtag lashkars of the past. Amanullah's commanders could muster some 50,000 regular troops, organized into 78 battalions of infantry, 21 regiments of cavalry and around 560 breech- and muzzle-loading guns. Amanullah's real strength rested in the vast hordes of Pashtun tribesmen who could be incited to rally to his standard. In December 1879, as many as 60,000 of these fighters had assembled for the siege of Roberts's Sherpur garrison. In the current conflict, the emir hoped to raise a lashkar of up to 30,000 in the Khyber region and perhaps as many as 80,000 on the Frontier as a whole.

Years of gun-running through the Persian Gulf, the capture of weapons from British stores and the militias and the manufacture of unreliable though deadly homemade firearms had now placed almost 300,000 modern and black powder rifles in tribal hands. The great advantage of enlisting the border tribes was their mobility and short supply lines. If the men ran out of provisions or ammunition, they simply returned home to gather more. In the same way, the great majority of today's Taliban fighters operate within a five-mile radius of their villages. Amanullah maintained in the Kabul arsenal 15,000 small-bore weapons and more than 400,000 Martini Henry rifles, capable of firing 12 rounds per minute and most of them supplied by the British Government. It was a far cry from the unruly fanatics who once made up Afghanistan's fighting force, but in spite of efforts to organize the army along Western lines, 'The Afghan regulars lacked training, and could not be considered as first class troops, although their courage and endurance were undoubted.'[13]

When war was declared, the British forces in India were well below their authorized strength. Large numbers of troops had been demobilized and sent home after the First World War and these were not replaced. Eight regular infantry and two cavalry regiments had been left in India during the Great War and the Indian units were temporarily short of effectives. Britain was fortunate in that a significant number of soldiers awaiting homeward-bound transport from the Middle East had been detained in India owing to the shortage of shipping. Many of these officers and men were called up to meet the Afghan crisis, but mobilizing them for active service was not a cut-and-dried affair. There was a lot of grumbling in the ranks over the demobilization orders. These were considered unfair in that instead of sending home the longest-serving troops first, farm hands and miners, deemed critical to rebuilding Britain's post-war economy, were given priority regardless of how long they had served.

When the news came that army units were being mobilized for combat duty on the North-West Frontier, the mood in the ranks verged on the mutinous. Meetings were held to discuss ways of avoiding active service on the Frontier and several Australian units refused point blank to have anything to do with it. In this case their commanders acquiesced rather than face a revolt in the ranks, and the men were shipped home. Those classed as Indian Army troops were not let off the hook and it wasn't until a number of perks were agreed, such as two weeks' pay in advance, that the troops were once again on the move.[14] The British could also count on the Frontier militias that had been raised following the inception of the Khyber Rifles in 1878. These numbered six corps deployed from Chitral in the north to South Waziristan. As events were to prove, in some cases the reliability of these native levies was compromised by their vulnerability to the fanatical preaching of Afghan-backed mullahs.

The forces that were mobilized to meet the Afghan incursion were divided into two battle groups. The North-West Frontier Force was under the command of General Sir Arthur Barrett, who held the office of general officer commanding Northern India and was a man of great Frontier experience. He became the senior officer on the ground throughout the war. Lieutenant General Richard Wapshare, a former master of the Bengal Foxhunt and a big game hunter with more than 50 tigers to his credit, was placed in charge of the Baluchistan Force. Wapshare boasted a somewhat less distinguished service record than Barrett's, having commanded at the blundered attempt to capture German East Africa (today Tanzania) in the First World War.

The first stage of hostilities began on 6 May on the Khyber front and this action involved regular Afghan units. By that time, three days after the invasion, the Afghans had placed three battalions of infantry with three guns at Bagh. Two hilltop positions about five miles north of Landi Kotal were also occupied by 350 Afghan troops and two guns. The Khyber Rifles garrison was thus effectively threatened from two sides, while a considerable force of Afghan infantry stood in reserve at Dakka, 13 miles east of Landi Kotal. There were 500 Khyber Rifles irregulars garrisoned at Landi Kotal, supported by two companies of Indian Infantry, a section of sappers and miners and another of mountain artillery. The troops crouching behind the fort's crenellated walls waited anxiously for the beat of Afghan war drums that preceded an advance, and they held out little hope of repelling the enemy once the attack began. The Afghans by now had in place at Bagh a powerful force of up to 8,000 regulars with at least a battery of mountain artillery and some quick-firing machine guns. Had they chosen to launch their offensive on that Tuesday evening, there was every probability of the hill tribes joining in the battle. It was nearly dark when the enemy massed in full at the walls, and the reserve troops were stationed at Peshawar, 22 miles away along a rough mountain road.

Happily for the defenders, the Afghan commander at Bagh, Mohammed Anwar Khan, let the opportunity slip. On 7 May, in conditions of great secrecy, one battalion from the Somerset Light Infantry was sent through the Khyber Pass to reinforce the Landi Kotal garrison. These troops were rushed through in a convoy of lorries covered with tarpaulins to conceal their loads from the tribesmen. The hooded lorries also deceived the military police, for three of them transported beer for the troops. Now it was the turn of the British and Indian forces to go on the offensive. Historian David Richards writes:

> At dawn on 9 May they began to move across the mile of barren ground which separated the fort form the Afghan position at Bagh. Its capture proved far from easy, for the troops could only progress on a narrow front over difficult sloping ground, around which covering fire was made almost impossible.[15]

The next day General George Crocker left Peshawar with a mixed force that reached Landi Kotal on 8 May, bringing the garrison up to brigade strength. Crocker spent the night mapping out his plan of attack on the Afghan regulars holding Bagh. The enemy was strung out along a strategic ridge about 2,000 yards long, commanding a position over the Khyber Rifles' fort. The Afghans were out in strength, with five battalions of infantry and six guns in position on the ridge. Crocker gave the order to move out at 4:45 a.m. under cover of darkness. As soon as the Afghans spotted what was afoot, they began laying down a heavy and accurate fire at the advancing Sikh and Gurkha troops. It quickly became apparent that a further advance on open ground would send the men to the slaughter.

After having recovered the water supply and taken a key hill position, the troops lacked the weight necessary to exploit their initial success. The men busied themselves scooping out shallow trenches under fire, in which to await reinforcements from Peshawar. There they remained for two days, whilst the situation at Landi Kotal steadily deteriorated and a picket of Khyber Rifles deserted with their weapons. These were the first of the levies to be wooed by the fanatical Afghan-backed mullahs, an act that led to the corps being disbanded for 20 years until the outbreak of the Second World War, when they were re-raised as the Afridi Battalion for service in the Middle East. Trouble in the militia ranks arose in other areas, notably in South Waziristan, where the locally-raised levies deserted after raiding the treasury and a store of ammunition.

Given the threat of desertions en masse, the Khyber Rifles were replaced by regulars in preparation for the second offensive on Bagh. Likewise, the South Waziristan Militia were disarmed, later coming back in their present incarnation as the South Waziristan Scouts. The attack began on the morning of 11 May, with the 2nd Infantry Brigade carrying the brunt of the assault, covered by fire from every available gun. After a sharp exchange of rifle and artillery fire, by mid-morning the Gurkha Rifles were in possession of Bagh. Their onslaught had been so sudden that those Afghans who managed to escape fled, leaving behind their artillery. The Gurkhas quickly captured the guns and bayoneted the gunners whilst they cleared the village of remaining enemy troops. The British and Indian units lost eight men in the Bagh offensive, a fraction of the more than 400 Afghans killed and wounded. As for fears that the tribesmen would flock by the thousands to their Afghan brethren, their minds were occupied with other matters.

> The rout was total and the tribesman that might have otherwise have been expected to counterattack in support of the Afghans decided against doing so, instead turning their efforts to looting the battlefield and gathering the arms and ammunition that the retreating Afghans had left behind.[16]

The battle tactics employed at Bagh were unremarkable in all respects but one, and this was to play a crucial role in determining the war's outcome. Lord Chelmsford telegraphed Secretary of State Montagu in London, giving an account of the engagement: 'Yesterday morning our advanced forces drove Afghans from their positions in neighbourhood of Landi Kotal, with co-operation of aeroplanes, who bombed Dakka with good effect.'[17] This was the first time that air power had been used in Frontier warfare. It was a tactic the tribesmen despised – they considered it unsporting – and more often than not it worked. For the first time, the fledgling Royal Air Force gave the British combat superiority over the enemy.

On 12 May, the first aircraft to be deployed in the war, a B.E.2 (Blériot Experimental),[18] swooped in to strafe the Afghans retiring across the Frontier after their defeat at Bagh. Three of these biplanes flew on to Dakka to hit the Afghan encampment, to the tribesmen's rage and the satisfaction of Roos-Keppel, who saw this as the way to hit the enemy's fighting spirit. Roos-Keppel, the scholar and fervent admirer of the Pashtuns, never left any doubts as to his true loyalty. In matters of Frontier defence, all other considerations were peripheral to preserving British rule. He decided not to let the Afghans off lightly and to prevent a counterattack by pursuing the enemy across the border. The RAF was once more called in to stage a bombing run on the Afghan garrison at Dakka. Roos-Keppel had written to Chelmsford's private secretary, urging a more vigorous use of air power: 'An attack on Dakka and possibly on Jalalabad from the air would not only take the heart out of the Afghans, but would give all those who are at present half-hearted a very good excuse for pulling out.'[19]

The aerial bombardment of Dakka sent the Afghans fleeing in terror from their stronghold, along the road to Jalalabad. On 12 May, Crocker took a battalion north to reconnoitre the Afghan position on two nearby hills. To their surprise the hills were found to be bristling with nearly 1,000 Afghan troops, ensconced behind sturdy stone breastworks. A brief skirmish erupted, following which Crocker marched his men back to camp, where he decided to ignore the Afghan position and instead make a bid to seize the more strategically important objective of Dakka. The town had been left virtually undefended and it offered the advantage of a staging ground for an advance on Jalalabad and Kabul. The next day, 13 May, the troops entered Dakka where they put up defences round the camp and prepared a landing strip for aeroplanes.

The next few days saw the Afghans return to fighting form in the Dakka area. There were sharp and hotly contested engagements to eject the Afghans from the hilltop strongholds that made Dakka untenable, and a number of British assaults went seriously wrong for a lack of proper artillery and ammunition. It was the arrival of reinforcements under General Sir Andrew Skeen, lauded by many as the outstanding Frontier soldier of his generation, that enabled the troops to regain the initiative. Skeen pounded the Afghan hilltop positions with howitzers ahead of an attack planned to begin at 2 p.m. But by 1 p.m. the Afghans had thrown in the towel and abandoned their makeshift fortresses, leaving behind five of their seven Krupp guns.

By mid-May, Afghan troops began infiltrating the territory of the people they held to be their closest Frontier allies, the Mohmands, in accordance with a plan to raise the tribes on both sides of the border against the British. The tribesmen's response was half-hearted, the Mohmands having had enough of fighting after a heavy defeat at the hands of British punitive expeditions four years earlier. The Mohmands were still suffering the effects of a long blockade and, to the Afghans' dismay, on this occasion some of them even volunteered to fight on the British side. The Afghans eventually managed to raise a small band of 4,000 tribesmen, some of them Mohmands and the rest from Bajaur to the north. The planned invasion of British territory rapidly turned into a non-event: British reinforcements were sent to defend Peshawar, the posts along the Mohmand blockade line were reinforced by infantry and machine guns, and the tribesmen ran out of provisions and dispersed back to their homes. The Afghan regulars, finding themselves lacking tribal support, returned to base.

Preparations were now underway for a British advance on Jalalabad, planned for 1 June. But the political authorities in London and Calcutta were already contemplating a cessation of hostilities. There was still confusion in official circles as to the exact nature of the conflict. Was this really a war, as it had been announced on both sides, or an outbreak of border clashes that could be ended with the withdrawal of the belligerents behind their respective lines? The latter school of thought was prompted by a letter from the Afghan commander-in-chief, Saleh Mohammed, to the Khyber political agent, seeking an armistice. Saleh launched into a tirade of righteous anger about the RAF 'throwing bombs from aeroplanes' and in the same breath declared a unilateral ceasefire pending talks between Amanullah and Chelmsford. The viceroy rightly dismissed the letter as 'insolent and so obviously a device to gain time'.[20]

Roos-Keppel was meanwhile busily translating letters from Amanullah, which had been intercepted by British-paid agents, and these told a different story. One was to a Bolshevist agitator in Peshawar, inviting him to join the emir in Kabul, and another promised to deliver troops and ammunition to an anti-British tribal leader. Roos-Keppel strongly recommended consigning the emir's peace overtures to the rubbish bin and to speed up the advance on Jalalabad. Accordingly, the 1st Division and 1st Cavalry Brigade were made ready for the march. The RAF sent B.E.2s over Jalalabad, the winter capital of Afghanistan, to soften up the town ahead of the advance along the Dakka-Jalalabad road. The planes dropped 1.8 tons of explosives in a single day and once more, in their time-honoured tradition, the Afridi tribesmen quartered in Jalalabad took advantage of the confusion to plunder the town before retiring to their villages, much to the disgust of the Afghan officials.

Dakka and Jalalabad had been subdued, but the Afghans had opened two other fronts when hostilities began, and these needed to be dealt with as well. Nadir Khan, the former commander-in-chief who had been replaced by Saleh Mohammed, was put in charge of the central command that posed a threat to Kurram as well as the Tochi Valley in North Waziristan. Prime Minister Abdul Qudus was sent to Kandahar, ignoring Quetta, which was well defended at division strength and could be quickly reinforced by a new wide-gauge railway through the Bolan Pass.

Qudus and his force remained largely out of action for the duration of the war, with the exception of one curious incident, the storming of the Spin Baldak fort, which lay on the Durand Line 40 miles south of Kandahar. On 29 May the eccentric General Wapshare decided that the capture of this fort would be a deterrent to the tribesmen and he surrounded the place with cavalry and two infantry brigades. Francis 'Ted' Hughes (later Brigadier), a 22-year-old subaltern in the 1st Gurkhas, recalled that prior to the attack senior officers, 'acting doubtless on the excellent principle that if you can't surprise the enemy it is better to surprise your own side than no one at all, supplied little or no information about the fort and its garrison'.[21] Hughes's regiment was ordered to take the south side of the fort's 15-foot-high outer wall in what was to be the last time the British Army used scaling ladders.

> The plan was to first place the scaling ladders in the ditch, so that the regiment could climb down one side and then up the

other. Then the men were meant to climb the wall, haul up the ladders, climb down, go through the ditch, and then climb the next wall. The Gurkhas were greatly diverted by this simple plan and declared that nothing like it had been seen since the siege of Jerusalem.[21]

Hughes says that everything was to be done in deathly silence:

> Indeed, the only sounds were the crashing of ammunition boxes and entrenching tools as the mules threw their loads, and the thudding of hooves as they bolted into the night. Every few seconds the air was split by the yells of some officer urging the men to greater silence or the despairing call of some NCO who had lost his section. A sound as of corrugated iron being dropped from a great height denoted that the scaling ladders were being loaded onto the carts: with these two exceptions, no one would have had an inkling that several thousand armed men were pressing forward to the fray.[22]

As things turned out, the scaling ladders were too short even for descending into the ditches, but after losing most of his men the garrison commander beat a hasty retreat to the hills with the remnants of his force. The British occupied the fort for a month, strengthened its defences, improved the water supply, and after the war they handed it back to the Afghans.

The Tochi region was the more alarming of the two theatres of operation under threat by Nadir Khan. The Kurram tribes were generally friendly to the British, whilst the belligerent Wazirs of Tochi could with little effort be incited to rise against the infidels. If the even more warlike Mahsuds of South Waziristan made common cause with their brethren to the north, the combined tribal force would prove a handful for the army. Fortunately Nadir Khan decided to direct his attack on Thal, bordering the Kurram River, where he arrived on 26 May with field guns strapped to the backs of elephants and mules. It was a bold move, for the road he had chosen for the march was believed to be impassable for field artillery, hence the British force was caught off guard. Thal fort came under heavy bombardment from the Afghan howitzers, and the RAF was called in to bomb the gun emplacements.

Relief arrived on 1 June under the command of Brigadier Reginald Dyer of Amritsar fame. Dyer marched his force nearly 30 miles in intense heat to scatter a lashkar of some 4,000 tribesmen. The following morning he pushed on relentlessly to take the main Afghan position on a hillside north-west of Thal. During this operation, Dyer was handed a letter from Nadir Khan stating that Emir Amanullah had instructed him to suspend hostilities. The no-quarter-given spirit that made Dyer a figure of contempt amongst Indians and raised him to iconic status in the ranks was reflected in the note he sent back: 'My guns will give an immediate reply.'[23] It soon became apparent that Nadir Khan and his forces were retiring with all possible speed. The RAF moved in to disperse a body of 400 tribesmen with bombs and Lewis-gun fire. Dyer sent armoured cars to harass the enemy's retreat. Preparations were now made to exploit the victory and follow Nadir Khan's retreating troops, but these plans were suddenly abandoned on 3 June when news was brought of a most unexpected event.

On 16 January 1919, Captain Robert 'Jock' Halley, reputed to be the smallest pilot in the RAF, had become the first aviator to fly from England to India, co-piloting his gigantic Handley-Page V/1500, dubbed 'Old Carthusian'. The flight took one month and four days. Halley was transferred to the Frontier and took part in operations during the war. At 3:30 a.m. on 24 May, Empire Day, Halley took off from Risalpur Airfield 28 miles east of Peshawar and turned his aircraft in the direction of Kabul. What was then the world's largest bomber, Halley's cumbersome four-engine biplane could manage a maximum speed of 90 knots and it took three hours for him to reach his destination. Once over Kabul, he lobbed some 20 bombs on the city, destroying its only ammunition factory and Emir Habibullah's tomb, and causing damage to Amanullah's palace and several other buildings.

Emir Amanullah emerged from the smouldering wreckage of his palace to survey the effects of this raid by a single RAF bomber. He contemplated the ruins of his father's tomb, he conferred with his ministers and field commanders and four days later, on 28 May, he sued for peace.

In a remarkable letter to Chelmsford, Amanullah strives to trivialize the war, stating that he did not wish 'to break our long-standing friendship or see enmity grow from bloodshed, and in proof of our good intentions we send this friendly letter, enclosing certified copies of orders sent to our commanders to stop movements of troops and cease from hostilities'.[24] The viceroy must have shaken his head in dismay to read Amanullah's impertinent claim that 'it was never the intention of my Government that our friendship should be severed'.[25] A curious gesture of friendship is this, that caused the mobilization of 340,000 troops and more than 1,750 British and Indian casualties, including some 500 dead in a cholera epidemic.

The Government of India granted an armistice on 3 June and a peace treaty was signed at Rawalpindi on 9 August 1919. The terms were simple and generous, in that no reparation was demanded of the Afghan Government or people, the British seeking only to guarantee the integrity of their border with India's neighbour. The Afghans were required to withdraw their troops 20 miles behind the Durand Line, whilst the British forces on Afghan soil would remain in place and aircraft would not be fired upon. Amanullah was to ensure the border tribes refrained from taking any aggressive action against the British Government. The Government would send a boundary commission to demarcate the undefined border west of the Khyber where the Afghan aggression took place, and the Afghans would accept whatever boundaries were set down by the British. In addition, the Government stripped Amanullah of his annual subsidy and prohibited him from importing arms and ammunition from India.

The treaty contained two other clauses of singular importance to both sides. The first of these required Amanullah to accept as valid and binding the international frontier established under the Durand Line agreement. Then, after the signing, Chelmsford telegraphed Montagu in London stating that the Afghans had inquired about Britain's future role in their country's external and internal matters. Amanullah said he was unwilling to renew the arrangement under which his father, Habibullah, had agreed to follow the advice of the British Government in matters concerning Afghanistan's foreign relations. Chelmsford at a stroke gave the emir great cause for jubilation and at the same time, got him off the hook with his disgruntled military: 'By the said Treaty and this letter, therefore, Afghanistan is left officially free and independent in its affairs, both internal and external. Furthermore, all previous treaties have been cancelled by this war.'[26]

So who had won this war, or month-long skirmish, as it was dubbed in some newspaper accounts? The official Government account states that despite the almost overwhelming difficulties that confronted the army in India at the outbreak of war, British arms were victorious. There was no doubt in Amanullah's mind as to who was the victor: he cheerfully distributed medals to his defeated generals and erected in Kabul a great column with a chained lion, representing Britain, at the base. This was meant to symbolize the Afghan victory at Thal, where the Afghan commander Nadir Shah had fled with his troops at full speed and with General Dyer in hot pursuit. On this the Government report comments, 'Many instances can be quoted in which nations, defeated in war, have laid claims to a larger degree of success than later history records, but it is seldom that a nation has gone so far as to claim victory when it has been defeated.'[27] The truth of the matter is that the Afghan regular troops were routed every time they were encountered and gained not a single success in the entire duration of the war.

Unfortunately, on balance there is some justification for Amanullah's victory celebrations. The emir was presented to his subjects as the Afghan leader who had defeated the mighty British Empire and regained his country's freedom. Moreover, shortly after the ratification of the Treaty of Rawalpindi, Amanullah scored another public relations victory when Britain's Great Game enemy, Russia, became the first country to recognize the newly-independent state of Afghanistan.[28] The emir now felt himself in a position of strength and began to resuscitate the social reforms initiated by his father. Some of these measures were quite revolutionary in nature. Amanullah advanced a liberal constitution that incorporated equal rights and individual freedoms, women were encouraged to remove their veils and Western dress was extended, along with education for both sexes.

In 1927, Amanullah embarked on a grand tour of Europe, wishing to observe at first hand the modern society he was so anxious to graft onto his country. The royal progress got off to an awkward start, when on the journey to India the emir pulled the emergency cord instead of the toilet chain as the train was going through a tunnel. Nevertheless, it was a successful visit, thanks in no small measure to the popularity of his glamorous wife Soraya Tarzi. Amanullah had the honour of being the first head of state to visit Germany since that country's defeat in the Great War. The Germans rolled out the presidential train for their distinguished visitor, with its heating apparatus unfortunately not functioning. The German newspapers reported that the royal couple and their entourage all appeared to be suffering from colds. Whilst in Berlin, the emir took delight in driving a U-Bahn train, which led to the class being known as the Amanullah-Wagen.

On his return to Kabul, Amanullah found his country in a less effusive state of mind about the radical reforms he sought to impose on this deeply conservative society. Amanullah came home to an uprising in Jalalabad, the country's fundamentalist heartland, which culminated in a march on the capital. Despite his efforts to rally the army, most of his troops deserted rather than resist and Amanullah was forced to flee from the same reactionary forces that had murdered his father. In 1928 he escaped over the Khyber Pass in his Rolls-Royce, clutching his favourite caged canary and, like several of his forebears, he went into exile in British India. The emir and his family eventually settled in Italy, having recalled that on his visit to Rome King Victor Emmanuel III had invited him to consider the country as his own.

★ ★ ★

There is no doubt that British arms prevailed in each of the three Afghan wars, but to what extent could these conflicts be taken as British victories? The British were outfoxed at every turn by the Afghans, a people with centuries of experience in suffering any manner of hardship to protect their independence and traditions. The Army of Retribution exacted its ruthless vengeance on the Afghans on its terrible march through the country in 1842. But this had been a war waged to effect regime change, in essence to oust an emir deemed to be hostile to Britain and replace him with an acquiescent British protégé. This policy resulted in abject failure, for no sooner had the Army of Retribution departed Afghanistan than the deposed ruler was returned to power and the protégé lay dead in a roadside ditch. And nothing could erase the reality of 16,000 corpses heaped in the snow on the road from Kabul to Jalalabad.

Thirty-six years later, a British invasion force once more marched out of Afghanistan, confident of having achieved its objective. A British Resident was imposed on Kabul to counter the perceived threat of Russian intrigues in Afghanistan. Within weeks, the Resident and his entire escort had been slaughtered and the army was on its way back to Afghanistan, this time extracting from Kabul control of the country's foreign relations and effectively turning it into a buffer state.

Another four decades were to pass before Britain once more clashed with its truculent neighbour, in the inglorious affair of the Third Afghan War. The supposed gains achieved by taking control of the country's foreign relations in 1878 were summarily relinquished 41 years later under the terms of the Treaty of Rawalpindi. The Russian threat that was meant to be neutralized by denying the Afghan Government relations with any foreign power except Britain turned into a reality when Russia became the first country to recognize the newly-independent state of Afghanistan and began pouring in material and economic assistance shortly after the 1919 war.

The Afghans can be beaten into submission, but never domination. The notion of a permanent, or even long-term occupation of the country is an illusion. The armies of every foreign conqueror throughout history have come to grief on Afghan soil. The Afghans are capable of enduring extreme hardship and sacrifice, including death, and they will fight with a savagery barely conceivable to a Western soldier, to defend their homeland. But it must be understood that the Afghan does not regard his 'homeland' as a nation state. Afghanistan is a conglomerate of martial tribes, dominated by the most warlike of them all, the Pashtuns, over which the emirs of the past and politician leaders of the present exercise but notional authority. An invasion by a foreign army is treated as a personal attack on each individual, his family and his immediate community. It is instructive to consider that the vast majority of insurgents who have joined the ranks of the resurrected Taliban fight within five miles of their villages.

This is not to suggest that the Afghans should be dismissed as intractable savages, a view that was all too prevalent in official circles in the days of the Raj – far from it. Britain's three Afghan wars, and distinctly the first one, demonstrate quite clearly the compulsory rules of engagement for dealing with the Afghans. At the risk of over-simplifying, these can be reduced to two: negotiate from a position of strength and scrupulously abide by one's word. One of the most amusing features of a tribal polo match is that the onlookers will invariably cheer the winning team. The favoured side is the one that scores the most goals. This principle can be transposed to the battlefield.

The British met with disaster the moment their vulnerability was exposed, be it by failing to take a vigorous stand against the Afghan rebels in 1841, or by leaving a British Resident virtually unprotected in Kabul in 1878. Most recently, we have witnessed the extraordinary sight of defeated Taliban fighters cheerfully defecting to their sworn enemy, the Northern Alliance, after the US invasion of 2001. It would be foolish to expect the Afghans to give their support to the Western powers' objectives of political and social reform without first proving supremacy in the conflict with the Taliban.

Then there is the issue of fair dealing. All was well whilst British India was doling out 'subsidies' to the Pashtuns to keep the roads and passes safe for the passage of troops and goods. The moment the tribal chiefs were summoned to Macnaghten's residence to be told that these allowances were to be halved, the Kabul garrison's doom was sealed. An Afghan will never forgive what he interprets as a betrayal, which is equivalent to an affront to his honour. Pashtunwali, the Pashtun code of honour, takes precedence over all other aspects of life, including Islam. Pashtunwali is composed of various binding precepts, which include hospitality to strangers, asylum to fugitives and forgiveness to enemies who offer their submission. Most important of all these principles is that of *badal*, or revenge. The Pashtun is bound to pursue and annihilate an enemy who has offended his or his family's honour. The Afghan people have very long memories and numerous blood feuds have been known to last for generations.

When this author began researching the history of Britain's three Afghan wars, it was not with the intention of drawing parallels with the current foreign political and military embroilment in that country. But the deeper one goes into the history of British involvement in Afghanistan, the more apparent it becomes that we do not always learn from our mistakes. Britain was not the last global power to encounter disaster in Afghanistan. No sooner had the Third Afghan War ended, the country began a 60-year roller-coaster ride that almost inevitably led to the Soviet invasion of 1979. Firstly, a rather objectionable Tajik named Kalakani seized the throne and ruled for only a few months. A Tajik had never sat on the throne in Kabul and the Pashtuns were simply not having it. Kalanaki was overthrown and killed by a Pashtun general, Mohammed Nadir Khan, who returned Afghanistan to orthodox Islamic rule. This worked for a while but in 1933 a student shot Nadir Khan dead in revenge for alleged atrocities committed against the Hazara minority.

Nadir's son Zahir Shah took over and this marked Afghanistan's coming-of-age. Under his leadership from 1933 to 1973, Afghanistan had a serious flirtation with liberal reform. Zahir Shah gave the country an enlightened constitution under a partially-elected legislature, and Afghanistan joined the League of Nations and signed a co-operation agreement with the US. The country later became a member of the United Nations and was a signatory to the Geneva Convention. In the cities, at least, Afghanistan began to recognize women as equal members of society. But the economy faltered as the country experimented with Western-style democracy, and in this state of crisis, in 1973, the king's cousin Mohammed Daoud, with the backing of leftist army officers, ousted Zahir, proclaimed a republic and turned to the Soviets for military aid to counter the US, which was actively arming Pakistan and Iran.

However, Daoud soon disappointed the military when he began to open negotiations with Iran, India and the US. The Afghan Army saw the Soviet Union as Afghanistan's most promising patron, hence in 1978 Daoud was ousted and then assassinated in another coup, this time bringing in a Soviet-backed regime under Mohammed Taraki, the founder of the People's Democratic Party of Afghanistan (PDPA), a political grouping that was communist in all but name. Afghanistan was now firmly locked into the Soviet camp, a development that was largely ignored by the US. Washington always acknowledged Afghanistan to be the Soviets' backyard and strategically less important than other countries of the region, such as Iran and Pakistan. Moreover, the US focus was on Tehran, where trouble was brewing that threatened to disrupt oil supplies from the Persian Gulf.

As is almost always the case with a revolutionary Marxist regime, radical economic re-engineering and land reform brought Afghanistan to its knees. The Government put on its agenda such inflammatory matters as equal rights for women and the redistribution of land. This was certain to trigger widespread hostility in the conservative rural community and the Russians knew it, so they urged Taraki to slow the pace. The Soviets feared that by moving too fast on these provocative issues, the communist regime would falter and stir up rebellion amongst the fundamentalist elements in the armed forces. They were right. Revolts broke out across the country and Taraki retaliated with mass arrests and executions. This was followed by a wave of political chaos, highlighted by the kidnapping and murder of the US ambassador and around 100 Soviet advisers and their families in Herat. Afghanistan was engulfed in widespread unrest and faced a popular uprising against the government, while parts of the army went over to the fledgling rebel movement.

Taraki was a committed Marxist but he backed a more independent line for Afghanistan than Babrak Karmal, who was a co-founder of the PDPA. Meanwhile a fight broke out between Taraki and his deputy, Hafizullah Amin, which only pushed the country deeper into mayhem. Amin had Taraki killed, allegedly by smothering him with a pillow. The Soviets were alarmed by the increasing disorder and also by President Jimmy Carter's decision to supply arms to the anti-communist rebels, though most of these weapons consisted of old Lee Enfield rifles. Alarm bells began ringing in earnest in the Kremlin.

With Pakistan the key US partner in the region and India closely allied with China, the Soviets were determined to keep Afghanistan under their wing. For a starter, the Russians needed to get rid of the loose cannon who was stirring up the anarchy. Amin was denounced as a CIA spy and was killed in murky circumstances. Babrak Karmal was brought back from exile in Moscow. The Soviets thought they now had a trustworthy agent in Kabul and in December of 1979 they began pouring military hardware into the country. For around seven years the Soviets could claim control of Afghanistan's major cities, but the countryside remained in the hands of the mujahedin insurgents, who were mercilessly carpet-bombed by Russian helicopters, against which the Afghans had no effective tactical weapon.

The turning point came in September 1986 when Soviet air supremacy was broken by US Stinger missiles. Within one year the insurgency had downed 270 Soviet aircraft, thus in 1989 the Soviets withdrew, fighting their way home just as the British had done more than a century before.

In keeping with historical precedent, the Afghans moved seamlessly from waging war against a foreign invader to warring amongst themselves. The Soviets left behind a society in ruins, with a hated communist government in power (which was to outlive the Soviet Union by three years) and a fragmented insurgency under the command of regional warlords. Then in the summer of 1994 a new group calling itself the Taliban emerged in Kandahar and by 1999 this regime of religious zealots had gained control of 90 per cent of the country. The rest is history familiar to everyone: the September 2001 attacks on the US, another foreign invasion of Afghanistan, the ousting and re-emergence of the Taliban and, at the time of writing, a desperate search for a glimmer of light at the end of the tunnel. By treating Afghanistan exclusively as a military issue, ignoring the country's social and economic problems, Britain threw away an opportunity to cultivate a fruitful and productive relationship with its neighbour, similar to the one it developed with India itself in the days of the Raj. It remains to be seen if the lessons of history will be applied to the current conflict in Afghanistan.

Notes

Introduction

1 Letters courtesy of William Eaton's great-great nephew, John Barstow.

Chapter 1

1 Elphinstone, Mountstuart, *An Account of the Kingdom of Caubul*, 1815 (reprint), Vol. I, p. 58.
2 Ibid., p. 59.
3 Ibid., p. 37.
4 Sykes, Percy, *A History of Afghanistan*, Cosmo Publications reprint, New Delhi, 2008, Vol. II, p. 388.
5 James, Lawrence, *Raj: the Making of British India*, Abacus, London, 1998, p. 63.
6 Campbell, George (Duke of Argyll), *The Afghan Question*, Straham & Co., London, 1880, p. 2.
7 The Maratha caste of Hindus consists mainly of warrior-landowners. Their vast empire declined gradually after a final defeat by the East India Company's army in 1761.
8 Snesarev, Andrei Evegenevich, *India as the Main Factor in the Central Asian Question*, St Petersburg, 1906, p. 173.
9 Mersey, Viscount Edward, *The Viceroys and Governors General of India*, John Murray, London, 1949, p. 39.
10 Habberton, William, *Anglo-Russian Relations concerning Afghanistan 1837–1907*, Illinois Studies in the Social Sciences, University of Illinois, 1937, p. 9.
11 Eden, Emily, *Up the Country*, Oxford University Press, Oxford, 1930, p. 225.
12 Ibid., p. 57.
13 Kaye, John William, *History of the War in Afghanistan*, W.H. Allen & Co., London, 1890, Vol. I, p. 369.
14 *Correspondence relating to Persia and Afghanistan*, Parliamentary Papers, 1839, No. 4, p. 4.
15 Pottinger, George, *The Afghan Connection*, Scottish Academic Press, Edinburgh, 1976, p. 29.

16   The Shia sect of Islam maintains that sovereignty resides in the Prophet Mohammed's descendants, starting with his son-in-law Ali. The Sunnis recognize the authority of the caliphs, starting with Medina and subsequently the Ummayads. The Shias were in the minority until the sixteenth century when the Iranian Safavid dynasty made it the official faith of their empire.
17   Ibid., p. 5.
18   Macrory, Patrick, *Signal Catastrophe*, Hodder and Stoughton, London, 1966, p. 270.
19   Shuja and Dost Mohammed belonged to rival branches of the Durrani lineage, the Pashtun dynasty that in the eighteenth century carved out an Afghan empire stretching from Persia to Kashmir. The Afghan Pashtuns are divided into two main groupings, the Durranis to the west and the Ghilzais, who occupy the eastern hills to the border with Pakistan. Shuja was of the Saddozai line, while Dost Mohammed was a Barakzai. In 1803, Shuja deposed his elder brother Shah Mahmud and ruled as Afghanistan's fifth emir until 1809, the year he concluded an alliance with Britain, when he was overthrown by his predecessor and fled into exile. Dost Mohammed came to power in 1834, to establish the Barakzai dynasty that ruled for nearly a century.
20   Pottinger, op. cit., p. 27.
21   Kaye, op. cit.,Vol. I, p. 218.
22   Ibid., p. 276.
23   Ibid., p. 310.
24   *Correspondence relating to Persia and Afghanistan*, Parliamentary Papers, 1839, No. 6, p. 6.
25   Ibid., No. 5, p. 7.
26   Ibid., No. 5, p. 25.
27   Ibid., No. 6, p. 7.
28   Ibid., No. 6, p. 8.
29   'Writers and scholars interested in Slavic and Asian history, languages and culture visited Orenburg frequently in the 1820s and 1830s, andVikevitch was encouraged to pursue his own studies of languages and literature.' Ingle, Harold, *Nesselrode and the Russian Rapprochement with Britain*, University of California Press, Berkeley, 1976, pp. 79–80.
30   *Correspondence relating to Persia and Afghanistan*, Parliamentary Papers, 1839, No. 6, p. 12.
31   Ibid., No. 6, p. 29.
32   Masson, Charles, *Narrative of Various Journeys in Balochistan, Afghanistan and the Panjab*, S. & J. Bentley, London, 1842,Vol. III, pp. 478–9.
33   *Indian Papers: Correspondence relating to Afghanistan*, IOR/V/4/1839/40, p. 47.
34   *Correspondence relating to Persia and Afghanistan*, Parliamentary Papers, p. 82.
35   Ibid., p. 176.
36   Ibid., p. 155.

Chapter 2

1. The president of the Board of Control was the ultimate authority in the Government of India. He was empowered by the East India Company's Secret Committee of the Court of Directors. The influence of the latter body was purely of a subordinate character, entirely dependent upon the president. The Company lost all its administrative powers after the 1857 Mutiny, when its Indian possessions, including its armed forces, were taken over by the Crown under the Government of India Act 1858.
2. *Broughton Papers*, Palmerston to Hobhouse, 46915, 18 October 1838, fos. 131–2.
3. Norris, J. A., *The First Afghan War*, Cambridge University Press, Cambridge, 1967, p. 215.
4. Kaye, op. cit., Vol. I, p. 129.
5. In a classic episode of Oriental horse-trading, Shuja had asked for three months' allowance in advance. He quickly amended this to six months in the hope of bettering his original request, and the Government settled on four months, amounting to 16,000 rupees, or £1,600 at that time.
6. Ranjit Singh also set out in the treaty an elaborate list of gifts that were to be exchanged between the two sovereigns. One clause prohibited the slaughter of cattle, but after some tough bargaining the Sikh king agreed to rescind this demand.
7. Kaye, op cit., Vol. 1, p. 131.
8. Ibid., p. 312.
9. Durand, Henry Marion, *The First Afghan War and Its Causes*, Longmans, Green & Co., London, 1879, p. 81.
10. *The Cambridge History of British Foreign Policy*, Cambridge University Press, Cambridge, 1923, Vol. II, p. 205.
11. *Correspondence with Lord Palmerston relative to the Late Sir Alexander Burnes*, British Library, Asian and African Studies, OIOC 1608/2252.
12. Sykes, op. cit., Vol. II, p. 1.
13. *Indian Papers: Correspondence relating to Afghanistan*, Treaty with Ranjit Singh and Shah Shuja ul Mulk, IOR/V/4/1839/40, p. 2.
14. 'Durbar' is a Persian word meaning the shah's noble court. *Hobson Jobson*, the delightful classic reference work for South Asian terminology, defines it as 'a Court or Levee' and also 'the Executive Government of a National State'. Under the British Raj, the word came to be applied to the place for a great ceremonial gathering.
15. Kaye, op. cit., Vol. I, p. 389.
16. Eden, op. cit., p. 208.
17. *Indian Papers: Correspondence relating to Afghanistan*, IOR/V/4/1840/Vol. 37, No. 1.
18. Kaye, John William, *The Life and Correspondence of Charles, Lord Metcalfe*, London, 1854.
19. Macrory, Patrick, *Signal Catastrophe*, Hodder & Stoughton, London, 1966, p. 84.

20 *Indian Papers: Correspondence relating to Afghanistan*, IOR/V/4/1840/Vol. 37, p. 7.
21 *Frontier and Overseas Expeditions from India*, Chief of Staff Army Headquarters, India, 1910, Vol. III, p. 307.
22 Kaye, op. cit., Vol. I, pp. 431–2, quoting Macnaghten's unpublished correspondence.
23 Ibid., p. 438.
24 Feringhee is a disparaging term, of Persian origin, for Europeans. Mountstuart Elphinstone was told on his journey that to call an Englishman a Feringhee was a positive affront.
25 *The Times*, 25 June 1842, p. 6.
26 The Kuzzilbash were the ancestors of immigrant mercenaries from Persia. The British were wary of this powerful Kabul clan's loyalties. Their apprehension was borne out when the Kuzzilbash threw in their lot with the insurgents who drove the army from Kabul.
27 The Dogras are an Indo-Aryan ethnic group, originally from the Jammu region of Kashmir. After the annexation of the Punjab by British India in 1849, they were recruited into the Indian Army and fought with great valour, earning two Victoria Crosses and 44 Military Crosses.
28 'Ghazi' stems from an Arabic word meaning 'to raid'. The Ghazis were religious zealots who depended upon plunder for their livelihood. They were chiefly responsible for the slaughter of the British garrison on the retreat from Kabul.
29 Kaye, op. cit., Vol. I, p. 461.
30 *Indian Papers: Correspondence relating to Afghanistan*, IOR/V/4/1840/Vol. 37, p. 12.
31 The renowned authority on Pashtuns, Olaf Caroe, who served as Governor of the North-West Frontier Province in 1946, had this to say on the tribesmen's character: 'Loyalty, to the point of love, will go to a forceful character. Win their devotion and these men will serve a leader to the death.' (Olaf Caroe, *The Pathans*, Macmillan & Co., London, 1958, p. 433.) The converse is equally true, in that any sign of weakness is taken as a signal to turn on one's master. Hence Dost Mohammed was neither surprised nor vexed when his mutinous troops broke into his tent and helped themselves to every item of value they could lay their hands on, right under the emir's sorrowful gaze.
32 Caroe, op. cit., p. 21.
33 Forbes, Archibald, *The Afghan Wars, 1839–1842 and 1878–1880*, Seeley & Co., London, 1892, p. 30.
34 Kaye, op. cit., Vol. I, p. 478.
35 Durand, op. cit., p. 187.
36 Sykes, op. cit., Vol. II, p. 13.
37 Fraser-Tytler, William Kerr *Afghanistan: A Study of Political Developments in Central and Southern Asia*, Oxford University Press, Oxford, 1950, p. 113.
38 Eyre, Vincent, *Journal of an Afghan Prisoner*, Routledge & Kegan Paul, London, 1976, p. 30.

39 Ibid.
40 MSS Eur B. 199. Marginal comments in a copy of Eyre, op. cit.
41 Kaye, op. cit., Vol. II, p. 5.
42 Moon, Penderel, *The British Conquest and Dominion of India*, Gerald Duckworth & Co., London, 1989, p. 521.

Chapter 3

1 British India at that time comprised two colonial regions. The Bengal Presidency consisted of present-day West Bengal and Bangladesh, as well as the states of Assam, Bihar, Meghalaya, Orissa and Tripura. Several princely states were later added to its jurisdiction, including the North-West Frontier Province, Punjab and Burma. The Madras Presidency was the other province directly under East India Company control. It encompassed much of southern India, which included Tamil Nadu, the Malabar region of North Kerala, parts of Andhra Pradesh and other regional districts.
2 Taken from the Old Testament (1 Kings 4:25): 'And Judah and Israel dwelt in safety, from Dan even to Beersheba, every man under his vine and under his fig tree, all the days of Solomon', as quoted in Kaye, op. cit., Vol. II, p. 137.
3 Fortescue, John William, *A History of the British Army*, Macmillan & Co., London, 1930, Vol. XII, pp. 164–5.
4 Fraser-Tytler, op. cit., p. 116.
5 Norris, op. cit., p. 338.
6 The nickname 'Iron Duke' comes only indirectly from Arthur Wellesley's forceful character. He was given that epithet because of the iron shutters he had fixed to the windows of his home to protect himself from parliamentary reform advocates during his term as prime minister.
7 Pottinger, George and Macrory, Patrick, *The Ten Rupee Jezail*, Michael Russell, Norwich, 1993, p. 197.
8 Macrory, op. cit., p. 136.
9 Fortescue, op. cit., p. 165.
10 Kaye, op. cit., Vol II., p. 146.
11 Broadfoot, William, *The Career of Major George Broadfoot*, John Murray, London, 1888, p. 34.
12 The Pashtuns of Afghanistan, who comprise about 42 per cent of the population, consider themselves the ruling class and superior in every respect to the less numerous groupings of Tajiks, Hazaras, Uzbeks and other smaller tribes. The country's Pashtuns are divided between the Abdalis, or Durranis as they came to be known in the eighteenth century, of the west, and the Ghilzais, who inhabit the rugged hills of the east that command the passes to India.
13 IOR, *The War in Afghanistan*, 9057, AA2, 1842, p. 13.
14 Forbes, op. cit., p. 74.
15 Ferrier, Joseph Pierre, *History of the Afghans*, John Murray, London, 1858, p. 343.

16 Kaye, op. cit., Vol. II, p. 410.
17 Broadfoot, op. cit., p. 40.
18 Kaye, op. cit., Vol. II, p. 206.
19 Quoted in ibid., p. 209.
20 Omrani, Bijan, 'Will We Make it to Jalalabad?', *Royal Society for Asian Affairs Journal*, London, Vol. XXXVII, No. II, p. 171.
21 Sale, Florentia, *A Journal of the First Afghan War*, Oxford University Press, Oxford, 1969, p. 16.
22 Gleig, George, *Sale's Brigade in Afghanistan*, John Murray, London, 1846, p. 125.
23 Quoted in Kaye, op. cit., Vol. II, p. 224.
24 Eyre, Vincent, *The Military Operations at Kabul*, John Murray, London, 1843, pp. 89–90.
25 Ibid., p. 91.
26 Durand, op. cit., p. 356.
27 Kaye, op. cit., Vol. II, p. 244.
28 The troops' red uniforms provided ideal targets for the Afghan skirmishers. In 1846, a few months before the Mutiny, Lieutenant Harry Lumsden raised a corps of irregulars to gather intelligence on tribal movements and act as guides to the troops in the field. One of Lumsden's first measures was to have his men abandon the tight-fitting scarlet uniforms worn by Indian Army troops. In their place, the Guides were to adopt a style of dress that blended in with the landscape. In doing so, Lumsden set a precedent in military history by outfitting his men in tunics and puttees of a new colour called *khaki*, an Urdu word of Persian origin meaning 'dusty' or 'dust coloured'.
29 Eyre, op. cit., p. 101.
30 Omrani, op. cit., p. 170.
31 Sale, op. cit., p. 82.
32 *Indian Papers: Correspondence relating to Afghanistan*, IOR/V/4/1839/40 p. 103.
33 Sale, op. cit., p. 85.
34 Sykes, op. cit., Vol. II, appendix B, p. 350.
35 Fortescue, op. cit., p. 231.
36 Ibid.
37 Sale, op. cit., p. 100.
38 Eyre, op. cit., p. 159.
39 Sale, op. cit., p. 104.
40 Kaye, op. cit., Vol. II, p. 385.

Chapter 4

1 Sale, op. cit., p. 98.
2 Gleig, op. cit., p. 137.

3 Eyre, op. cit., p. 241; emphasis in original.
4 *The Times*, 5 May 1842, p. 7.
5 Kaye, op. cit., Vol. III, p. 4.
6 Auckland to Hobhouse, 18 February 1842, IOR Auckland 37707 fos. 187–8.
7 Ferrier, op. cit., p. 363.
8 His wife, Jane Elizabeth Digby, was an artist of great beauty and somewhat casual virtue. She afterwards became mistress of the king of Bavaria and later married his prime minister, Baron von Venningen, who committed suicide when she deserted him. Her third husband was an Arab general, with whom she lived in Damascus until her death in 1881.
9 Fortescue, op. cit., p. 263.
10 Norris, op. cit., p. 391.
11 Pollock's appointment had provoked some friction between Ellenborough and Nicolls. The commander-in-chief in India had wanted Nott to take over as commander in Afghanistan, but was overruled by the new governor general.
12 Ellenborough to Pollock, 1 March 1842, PRO 30.12.98.
13 Greenwood, John, *The Campaign in Afghanistan*, Nonsuch Publishing reprint, Stroud, 2005, p. 109.
14 *Journal of the Society for Army Historical Research*, 1967, Vol. 45, p. 223. Cumming was killed at Jamrud on the advance into the Khyber Pass on 5 April 1842, his 23rd birthday. The day before his death he ended a letter to his father with the prophetic words, 'I find it just post time, and I must close now, for I know not where I may be tomorrow. Accept, my dearest and best of fathers, a soldier's farewell.'
15 Fortescue, op. cit., p. 264.
16 There is a British cemetery at Ali Masjid, with the graves of soldiers killed in the Second Afghan War. The valley walls bear insignia of British regiments that served here, including the Gordon Highlanders, the South Wales Borderers, and the Royal Sussex, Cheshire and Dorset regiments.
17 Greenwood, op. cit., p. 114.
18 Ibid., p. 117.
19 Ibid.
20 Queen Victoria was so impressed by the regiment's valour that she ordered it should henceforth be known as 'Prince Albert's Regiment of Light Infantry', after her consort. The Queen also approved the change of colour of the regiment's uniform facings from yellow to royal blue. A special campaign medal was struck to commemorate the siege. In Parliament, Prime Minister Peel referred to the 13th as the 'Illustrious Garrison'.
21 Greenwood, op. cit., p. 164.
22 Hamilton, Major Alexander, 'The First Afghan War', *Journal of the United Service Institution of India*, July 1908, Vol. XXXVII, No. 172, p. 323.
23 *The Times*, 27 January 1843, p. 4.
24 Royal Archives RA/VIC/N12/17.
25 Ibid.
26 There is debate as to how many people had been taken prisoner by Akbar

in Kabul and on the retreat from Kabul. The best estimate would put the figure at around 100, comprising an estimated 22 officers, 37 other ranks, 19 wives and 22 children.

27 Afghan fighters are prone to change sides with the same facility with which they change their socks. Witness the many ranks of Taliban who cheerfully joined forces with the Northern Alliance in 2001 when the US launched its invasion of Afghanistan.

28 Lawrence, George, *Reminiscences of Forty-Three Years in India*, John Murray, London, 1874, p. 222.

29 Sale, op. cit., p. 157.

30 Mackenzie, Colin, *Storms and Sunshine of a Soldier's Life*, David Douglas, Edinburgh, 1884, p. 365.

31 Greenwood, op. cit., pp. 164–5.

32 McCaskill was killed three years later fighting the Sikhs at the Battle of Mudki. Brigadier Sale also fell mortally wounded in that engagement, which ended in a victory for the British, at a cost of 870 casualties.

33 The Proclamation consisted of a potted summary of the recent war, explaining that 'The Government of India directed its army to pass the Indus to expel from Afghanistan a chief *believed* to be hostile to British interests, and replace upon his throne a sovereign *represented* to be friendly to those interests, and popular with his former subjects' (emphasis in original). He then went on to inform his subjects, in what was almost a verbatim repetition of the hypocritical assurances given by Auckland to Dost Mohammed, that 'To force a sovereign upon a reluctant people, would be as inconsistent with the policy, as it is with the principles of the British Government.'

Chapter 5

1 Rawlinson, Henry, The Afghan Crisis, *Nineteenth Century*, No. XXII, Edinburgh, 1878, p. 3.

2 Akbar Khan is a highly revered figure in Afghanistan today. There is even a residential neighbourhood in Kabul named after him. It is the city's most affluent suburb and home to many embassies.

3 Afghan dynastic law prohibits the blind from occupying the throne.

4 Moon, op. cit., p. 565.

5 Ibid.

6 George Douglas Campbell, Duke of Argyll, served as secretary of state for India from 1868 to 1874 under the Gladstone ministry. The quote is taken from ibid. p. 2.

7 Spear, Percival, *The Oxford Modern History of India*, Oxford University Press, Oxford, 1965, p. 165.

8 The British had signed a treaty with the emirs of Sind in 1832, in which Calcutta agreed never to use the Indus for a military crossing. The 1838

invasion of Afghanistan through Sind constituted a violation of that treaty, but the Sind rulers acquiesced to avoid an attack on Hyderabad by the Army of the Indus. This time they signed two more treaties, paid a tribute to the British and provided the army with supplies. Troops took over Karachi and a British force was stationed there to make sure tolls were no longer collected on the Indus River.

9 The original of this telegram has never been found. It may sadly be apocryphal, though it was reproduced in an 1844 *Punch* cartoon.
10 *The Cambridge History of British Foreign Policy*, A. W. Ward and G. P. Gooch (eds), Cambridge University Press, Cambridge, 1923, Vol. II, p. 211.
11 The Rajputs are one of the major Hindu warrior groups of India, whose origins date from the seventh century AD.
12 For a highly-entertaining account of the Sikh dynastic intrigues, see George Macdonald Fraser's *Flashman and the Mountain of Light*.
13 Spear, op. cit., p. 173.
14 Fortescue, op. cit., p. 346.
15 Mason, Philip, *A Matter of Honour*, Jonathan Cape, London, 1974, pp. 228–9.
16 These two officers, of different nationalities, travelled together to Lahore to fight for the Sikhs. They are credited with having devised the successful strategy that brought about the Sikh victory over the Afghans at the battle of Nowshera in 1822, which resulted in the capture of Peshawar.
17 Fortescue, op. cit., p. 351.
18 Sale's widow, the indomitable Lady Florentia Sale, continued to reside in India on a handsome pension of £500 a year. In 1853, at the age of 63, she travelled to the Cape of Good Hope for health reasons and died a few days after her arrival.
19 Spear, op. cit., p. 174.
20 Churchill, Winston, *The Story of the Malakand Field Force*, Leo Cooper reprint, London, 2002, p. 230.
21 Allen, Charles, *Soldier Sahibs*, John Murray, London, 2000, p. 255.
22 Ibid., p. 283.

Chapter 6

1 Fraser-Tytler, op. cit., p. 123.
2 *Glasgow Herald*, 3 March 1943, p. 3.
3 Robson, Brian, *The Road to Kabul*, Spellmount, Stroud, 2007, p. 34.
4 Sykes, op. cit., Vol. II, p. 86.
5 Campbell, op. cit., p. 9.
6 Ibid.
7 Ibid. p. 3.
8 *The Times*, 29 January 1873, p. 8.
9 Not everyone in Government was obsessed with Russian expansion in Central Asia. There was some dissention over the prevailing alarm at the tsar's

adventurist policies, once colourfully expressed in the House of Commons by Sir Charles Wingfield MP in February 1873. Wingfield proclaimed that Russia's annexation of Khiva was 'in the interests of humanity' and he wished Russia every success in reducing 'this degraded population to order and civilisation'. *Hansard*, February 1873, Vol. 214.

10 Monypenny, William and George Buckle, *The Life of Benjamin Disraeli*, John Murray, London, 1929, Vol. II, p. 155.
11 Ibid.
12 Balfour, Betty, *The History of Lord Lytton's Indian Administration*, Longmans, Green & Co., London, 1899, pp. 2–3. Balfour was Lytton's daughter and had access to information recorded in his private papers, some of which was only made public in her book.
13 This exchange was never recorded and is quoted in Balfour, op. cit., pp. 35–6.
14 Balfour, op. cit., p. 37.
15 *Parliamentary Papers on Afghanistan*, IOR, L/PAR/2/88, p. 53.
16 Ibid., p. 90.
17 Ibid., p. 94.
18 *The Times*, 14 November 1878, p. 4.
19 Ibid., p. 105.
20 Britain was given the right to do this under a treaty signed with the Khan of Khelat in 1854.
21 *Parliamentary Papers on Afghanistan*, IOR, L/PAR/2/88, p. 170.
22 Ibid.
23 Dutt, Romesh, *India in the Victorian Age*, Kegan Paul, London, 1904, p. 428.
24 Chamberlain said that the destruction of Istalif had left him 'disgusted with myself, the world, and above all, with my cruel profession. In fact we are nothing but licensed assassins.' Quoted in Forrest, George, *Life of Field Marshal Sir Neville Chamberlain*, William Blackwood & Sons, Edinburgh, 1909, p. 149. Other distinguished members of this family included Prime Minister Arthur Neville Chamberlain (1937–40) and Neville Francis Chamberlain, army officer and the inventor of snooker.
25 Balfour, op. cit., p. 280.
26 Ibid., p. 285.
27 Ibid., p. 286.
28 *Parliamentary Papers on Afghanistan*, IOR, L/PAR/2/88, p. 199.
29 Ibid., p. 295.
30 Browne lost his left arm to an enemy sword during the Sepoy Mutiny, leaving him unable to control or draw his sword. He came up with the idea of wearing a second belt which went over his right shoulder and held the scabbard in just the spot he wanted. This would hook into a heavy leather belt with D-rings for attaching accessories. This soon became part of the standard uniform.
31 The total strength of the three columns marching on Afghanistan, according to General Roberts, was as follows:
    – Kandahar Field Force: 265 officers, 12,599 men, 78 guns

NOTES 243

- Kurram Field Force: 116 officers, 6,549 men, 18 guns
- Peshawar Valley Field Force: 325 officers, 15,854 men, 48 guns.
32 Letter from Lord Salisbury to Queen Victoria, 26 April 1899, Royal Archives, RA/VIC/A 75/61.
33 Roberts, Frederick, *Forty-One Years in India*, Macmillan & Co., London, 1914, p. 346.
34 It is a wonder that the troops were able to even inch forward, much less keep up a marching pace, given the enormous quantity of officers' baggage taken into Afghanistan and the army of camp followers required to carry it. One senior officer's journal records for his personal use 200 lbs' weight of tents, a six-foot bed, a folding table and chair, a tin basin for washing, three saucepans, a kettle, a stewing pan, a frying pan, a pewter teapot, two tea cups and saucers, two soup plates and dinner plates, enamelled iron, a luncheon basket with three knives, forks, spoons, two glasses and a dozen cases of whisky, 'so in a pinch I can give a dinner party or two'. The officer's retinue of servants and mounts consisted of a general man servant, three native grooms and three horses. (Quoted in Brooke, Henry, *Brigade Commander*, Leonaur reprint, 2008, pp. 14–15.)
35 Roberts, op. cit., p. 353.
36 Ibid., p. 355.
37 The Turis are the tribesmen of the Kurram Valley. Unlike their immediate neighbours and the vast majority of the Pashtuns, the Turis are Shia not Sunni Muslims. They suffered great depredations at the hands of the Sunnis and at one point petitioned the Government of India to take them under British protection.
38 Roberts, op. cit., p. 368.
39 Forbes, Archibald, *Britain in Afghanistan: The Second Afghan War, 1878–80*, Seeley & Co., London, 1892, p. 23.
40 Balfour, op. cit., p. 295.
41 Robson, op. cit., p. 107.
42 Ibid., p. 381.

Chapter 7

1 Roberts, op. cit., p. 383.
2 Robson, op. cit., *Road to Kabul*, p. 120.
3 Macgregor, Charles Metcalfe, *The Second Afghan War: Official Account*, John Murray, London, 1908, p. 184.
4 *Sowar* is the Persian word for 'horseman', or literally 'one who rides'. In the Indian Army a *sowar* was a native cavalry officer, or a mounted orderly. It was the rank equivalent to sepoy in the infantry. Cavalry troopers in the modern armies of India, Pakistan and Bangladesh still hold this rank.
5 Balfour, op. cit., p. 359.
6 Roberts, op. cit., p. 386.

7   Farwell, Byron, *Eminent Victorian Soldiers*, W.W. Norton & Co., New York, 1985, p. 167.
8   Roberts, op. cit., p. 395.
9   Ibid., p. 396.
10 The heliograph is a wireless solar telegraph that signals using Morse code flashes. It was extensively used by British troops on the North-West Frontier in campaigns against the Pashtuns, for relaying intelligence on enemy positions and as an SOS system. Some of Pakistan's Frontier Scouts still train with the heliograph.
11 Roberts, op. cit., p. 407.
12 *The Illustrated London News*, 11 October 1879, p. 34.
13 Roberts, op. cit., p. 407.
14 Ibid., pp. 410–11.
15 Farwell, op. cit., pp. 166–7.
16 Roberts had a keen awareness of the lessons of history. The Sherpur cantonment's four-and-a-half-mile perimeter was a drawback for its defence. Roberts remembered the disaster of Elphinstone's scattered force in 1841, but reasoned that the ability to keep all the troops together far outweighed the disadvantage of having to defend so long a line.
17 Hensman, Howard, *The Afghan War*, H. Allen & Co., London, 1881, p. 137.
18 Forbes, Archibald, *Britain in Afghanistan*, Leonaur, London, 2007, p. 75.
19 *Lashkar* is a Persian word meaning 'army' or 'military camp'. It has passed into Anglo-Indian terminology to describe a tribal fighting force and was commonly used in official military reports during the British occupation of the North-West Frontier.
20 Hensman, op. cit., p. 189.
21 Roberts, op. cit., p. 449.
22 Ibid., p. 453.
23 Ibid., p. 457.
24 Balfour, op. cit., p. 396.
25 Sykes, op. cit., Vol. II, p. 120.
26 Roberts, op. cit., p. 464.
27 Ibid., p. 465.
28 Frederick John Robinson, first Viscount Goderich, held office for just five months, from August 1827 to January 1828. His ministry was plagued by party feuding and resignations. Depressed by his wife's poor health and unable to control his obstinate colleagues, Goderich concluded that his Government was about to break up. On 8 January 1828 he reported the situation to King George IV. The final ignominy came when the king asked Goderich to take the necessary steps to arrange for his own replacement – a request that led Goderich to break down in tears and the king to pass him a handkerchief.
29 Mersey, op. cit., p. 100.
30 Balfour, op. cit., p. 434.
31 Ibid.
32 Roberts, op. cit., p. 468.

33 Doyle, Arthur Conan, *A Study in Scarlet*, Ward, Lock & Co., London, 1888, p. 3. The disaster at Maiwand also inspired Kipling to write 'That Day', a poem commemorating the last stand of the 66th Berkshire Regiment.
34 Hills, John, *The Bombay Field Force*, R. Brimley Johnson, London, 1900, p. 15.
35 Burrows's brigade was comprised of the following: E/B Battery, Royal Horse Artillery, 3rd Queen's Own (Bombay Cavalry), 3rd Sind Horse (Bombay Army), HM 66th of Foot, 1st Grenadiers (Bombay Army), 30th Bombay Native Infantry (Jacob's Rifles), 2nd Company Bombay Sappers and Miners. Yakub Khan commanded 16 Afghan regular infantry regiments, 32 guns, three regiments of regular cavalry plus 3,000 irregulars, 4,000 mounted irregulars, as well as every male from the surrounding countryside who was capable of bearing arms.
36 Hills, op. cit., p. 17.
37 *Nullah* is a Hindi word for 'ravine', usually a dry river course, which in Frontier warfare formed one of the toughest obstacles to troop movement. The Pashtuns, however, were adept at negotiating these *nullahs*, often using them as hiding places or as quick escape routes after staging a raid.
38 Hills, op. cit., pp. 29–30.
39 Quoted in *The Times*, 30 August 1880, p. 5.
40 Roberts, op. cit., p. 471.
41 Ibid., p. 485.

Chapter 8

1 Fraser-Tytler, op. cit., p. 165.
2 Ibid., p. 188.
3 Churchill, op. cit., p. 33.
4 The Third Afghan War 1919, Official Account, Army Headquarters, Calcutta, 1926, p. 11.
5 Sykes, op. cit., Vol. II, p. 268.
6 Fraser-Tytler, op. cit., p. 195.
7 The brutality of the Amritsar massacre was unlike anything British India had experienced before or after. The firing continued without interruption for a quarter of an hour. Dyer then immediately withdrew, leaving the dead, dying and wounded unattended. Further action was taken on 19 April when Dyer issued an order that all Indians passing along the street in which a British missionary had been murdered were to do so crawling on their hands and knees. The entire episode was a godsend to Gandhi and his supporters of the Quit India movement.
8 Jalazai, Musa Khan, The Foreign Policy of Afghanistan, Sang-e-Meel, Lahore, 2003, p. 69.
9 Robson, Brian, The Third Afghan War, Spellmount, Staplehurst, 2004, p. 12.
10 Papers regarding Hostilities with Afghanistan 1919, HMSO, London, 1919, p. 8.

11 There was one incident that illustrated the mutual respect prevailing between adversaries on the Frontier. Roos-Keppel, who commanded in the 1908 campaign to pacify the notorious Zakka Khel clan of the Afridi tribe, came forward to accept the submission of one of the maliks. 'Did we not fight well, sahib?' the greybeard asked, to which Roos-Keppel replied, 'I wouldn't have shaken hands with you unless you had.'
12 Fremont-Barnes, Gregory, The Anglo-Afghan Wars 1839–1919, Osprey Publishing, Oxford, 2009, p. 81.
13 The Third Afghan War 1919, Official Account, p. 23.
14 Diary of E. Holter, Imperial War Museum records, 81/9/1.
15 Richards, David, The Savage Frontier, Macmillan & Co., London, 1990, p. 260.
16 Wilkinson-Latham, Robert, Discovering Artillery, Shire Publishers, Oxford, 1998, p. 23.
17 Papers regarding Hostilities with Afghanistan 1919, p. 10.
18 The B.E.2 was a single-engine, two-seat biplane that was considered untrustworthy when it was brought into service on the North-West Frontier. The aircraft was underpowered, it was easy prey for the Germans in the Great War (when it was dubbed 'Fokker Fodder' by the British press), and the observer and gun had to be left behind when it was fully loaded with bombs.
19 Quoted in Heathcote, op. cit., p. 127.
20 Papers regarding Hostilities with Afghanistan 1919, p. 12.
21 Quoted from the Daily Telegraph obituary, 17 April 1987.
22 Ibid.
23 The Third Afghan War 1919, Official Account, p. 61.
24 Papers regarding Hostilities with Afghanistan 1919, p. 25.
25 Ibid.
26 Ibid., p. 36.
27 The Third Afghan War 1919, Official Account, p. 136.
28 This foreshadowed the future course of Afghan and indeed Cold War history, for the United States refused to recognize Afghanistan until 1934, thus giving the Bolshevists an early foothold in the country.

Bibliography

Abbreviations
HMSO His Majesty's Stationery Office
IOR India Office Records (British Library)
MSS Eur European Manuscripts (British Library)
OIOC Office of Indian and Oriental Collection (British Library)

Primary Sources
Auckland to Hobhouse, IOR 37707, fos. 187–8, 18 February 1842.
Broughton Papers, Palmerston to Hobhouse, 46915, 18 October 1838, fos. 131–2.
Correspondence relating to Persia and Afghanistan, Parliamentary Papers, London, 1890, No. 4, p. 4; ibid., No. 5, pp. 7, 25; ibid., No. 6, pp. 6, 7, 8.
Correspondence with Lord Palmerston relative to the Late Sir Alexander Burnes, OIOC 1608/2252.
Frontier and Overseas Expeditions from India, Chief of Staff Army Headquarters, India, 1910, Vol. III, p. 307.
Indian Papers: Correspondence relating to Afghanistan, IOR/V/4/1839/40, p. 2; IOR/V/4/1840/Vol. 37, pp. 7, 12.
Letter from Lord Salisbury to Queen Victoria, Royal Archives, 26 April 1899, RA/VIC/A 75/61.
MSS Eur B. 199. Marginal comments in a copy of Eyre, Vincent, *The Military Operations at Kabul*, John Murray, London, 1843.
Papers regarding Hostilities with Afghanistan 1919, HMSO, London, 1919, p. 8.
Parliamentary Papers on Afghanistan, IOR L/PAR/2/88, pp. 53, 170, 199, 295.
Royal Archives, RA/VIC/N12/17.
Sale, Florentia, *A Journal of the First Afghan War*, Oxford University Press reprint, Oxford, 1969.

The Third Afghan War 1919, Official Account, Army Headquarters, Calcutta, 1926.
The War in Afghanistan, IOR 9057, AA2, 1842, p. 13.

Newspapers and Journals

Anglo-Russian Relations concerning Afghanistan 1837–1907, Habberton, William, *Illinois Studies in the Social Sciences*, University of Illinois, 1937, p. 9.
Daily Telegraph, 17 April 1987, p. 44.
Glasgow Herald, 3 March 1943, p. 3.
Journal of the Society for Historical Research, Vol. 45, 1967, p. 223.
Journal of the United Service Institution of India, 'The First Afghan War', Hamilton, Alexander, July 1908, Vol. XXXVII, No. 172, p. 323.
Nineteenth Century, 'The Afghan Crisis', Rawlinson, Henry, No. XXII, Edinburgh, 1878, p. 3.
Royal Society for Asian Affairs Journal, 'Will We Make it to Jalalabad?', Omrani, Bijan, London, Vol. XXXVII, No. II, p. 171.
The Illustrated London News, 11 October 1879, p. 34.
The Times, 5 May 1842, p. 7; 27 January 1843, p. 4; 29 January 1873, p. 8; 14 November 1878, p. 4; 30 August 1880, p. 5.

Secondary Sources

Allen, Charles, *Soldier Sahibs*, John Murray, London, 2000.
Balfour, Betty, *The History of Lord Lytton's Indian Administration*, Longmans, Green & Co., London, 1899.
Barthorp, Michael, *Afghan Wars and the North-West Frontier 1839-1947*, Cassell & Co., London, 1982.
Broadfoot, William, *The Career of Major George Broadfoot*, John Murray, London, 1888.
Brooke, Henry, *Brigade Commander*, Leonaur reprint, London, 2008.
Campbell, George (Duke of Argyll), *The Afghan Question*, Straham & Co., London, 1880.
Caroe, Olaf, *The Pathans*, Macmillan & Co., London, 1958.
Churchill, Winston, *The Story of the Malakand Field Force*, Leo Cooper reprint, London, 2002.
Doyle, Arthur Conan, *A Study in Scarlet*, Ward, Lock & Co., London, 1888.
Durand, Henry, Marion, *The First Afghan War and Its Causes*, Longmans, Green & Co., London, 1879.
Dutt, Romesh, *India in the Victorian Age*, Kegal Paul, London, 1904.
Heathcote, Anthony, *The Afghan Wars*, Osprey, London, 1980.
Eden, Emily, *Up the Country*, Oxford University Press, Oxford, 1930.

Elphinstone, Mountstuart, *An Account of the Kingdom of Caubul*, 1815, Indus Publications reprint, 1992.
Eyre, Vincent, *The Military Operations at Kabul*, John Murray, London, 1843.
Farwell, Byron, *Eminent Victorian Soldiers*, W.W. Norton & Co., New York, 1985.
Ferrier, Joseph Pierre, *History of the Afghans*, John Murray, London, 1858.
Forbes, Archibald, *The Afghan Wars, 1839–1842 and 1878–1880*, Seeley & Co., London, 1892.
——, *Britain in Afghanistan: The Second Afghan War, 1878–80*, Leonaur, London 2007.
Forrest, George, *Life of Field Marshal Sir Neville Chamberlain*, Wm. Blackwood & Sons, Edinburgh, 1909.
Fortescue, John William, *A History of the British Army* (13 vols), Macmillan & Co., London, 1930.
Fraser-Tytler, William Kerr, *Afghanistan: A Study of Political Developments in Central and Southern Asia*, Oxford University Press, Oxford, 1950.
Fremont-Barnes, Gregory, *The Anglo-Afghan Wars 1839–1919*, Osprey Publishing, Oxford, 2009.
Gleig, George, *Sale's Brigade in Afghanistan*, John Murray, London, 1846.
Greenwood, John, *The Campaign in Afghanistan*, Nonsuch Publishing reprint, Stroud, 2005.
Hensman, Howard, *The Afghan War*, H. Allen & Co., London, 1881.
Hills, John, *The Bombay Field Force*, R. Brimley Johnson, London, 1900.
Ingle, Harold, *Nesselrode and the Russian Rapprochement with Britain*, University of California Press, Berkeley, 1976.
Jalazai, Musa Khan, *The Foreign Policy of Afghanistan*, Sang-e-Meel, Lahore, 2003.
Kaye, John William, *The Life and Correspondence of Charles, Lord Metcalfe*, London, 1854.
—— *History of the War in Afghanistan* (3 vols), W.H. Allen & Co., London 1890.
Lawrence, George, *Reminiscences of Forty-Three Years in India*, John Murray, London, 1874.
Lawrence, James, *Raj: The Making of British India*, Abacus, London, 1998.
Macgregor, Charles Metcalfe and Murray, John, *The Second Afghan War: Official Account*, London, 1908, p. 184.
Mackenzie, Colin, *Storms and Sunshine of a Soldier's Life*, David Douglas, Edinburgh, 1884.
Macrory, Patrick, *Signal Catastrophe*, Hodder and Stoughton, London, 1966.
Mason, Philip, *A Matter of Honour*, Jonathan Cape, London, 1974.

Masson, Charles, *Narrative of Various Journeys* (3 vols), S. & J. Bentley, London, 1842.
Mersey, Edward, *The Viceroys and Governors General of India*, John Murray, London, 1949.
Monypenny, William and Buckle, George, *The Life of Benjamin Disraeli*, John Murray, London, 1929.
Moon, Penderel, *The British Conquest and Dominion of India*, Gerald Duckworth & Co., London, 1989.
Norris, James, *The First Afghan War*, Cambridge University Press, Cambridge, 1967.
Pottinger, George, *The Afghan Connection*, Scottish Academic Press, Edinburgh, 1976.
Pottinger, George and Macrory, Patrick, *The Ten Rupee Jezail*, Michael Russell, Norwich, 1993.
Richards, David, *The Savage Frontier*, Macmillan & Co., London, 1990.
Roberts, Frederick, *Forty-One Years in India*, Macmillan & Co., London, 1914.
Robson, Brian, *Crisis on the Frontier*, Spellmount, Stroud, 2004.
——, *The Road to Kabul*, Spellmount, Stroud, 2007.
Snesarev, Andrei Evegenevich, *India as the Main Factor in the Central Asian Question*, St Petersburg, 1906.
Spear, Percival, *The Oxford Modern History of India*, Oxford University Press, Oxford, 1965.
Sykes, Percy, *A History of Afghanistan* (2 vols), Cosmo Publications reprint, New Delhi, 2008.
Ward, Adolphus William, Sir, and Gooch, George Peabody (eds), *The Cambridge History of British Foreign Policy*, Cambridge University Press, Cambridge, 1923, Vol. II, p. 211.
Wilkinson-Latham, Robert, *Discovering Artillery*, Shire Publishers, Oxford, 1998.

Index

Abdullah Jan 143, 146
Abdur Rahman 196, 199–205, 208, 212
Afghanistan
  Britain gains control of foreign
    policy xv, xviii, 226
  Britain returns control of foreign
    policy xv, 226
  Soviet invasion (1979) xviii, 228
  Soviet withdrawal (1989) 231
Afghans
  tribal customs xv
  Shuja on 41
  list of demands at Kabul 82
  massacre on the way to Jalalabad
    85–8
  barbarism towards sepoys 107
  personality xx, 227, 228
  Cavagnari's attitude 164
  Roberts' ruthless handling of 166–72
Afridis 54, 100, 102, 113, 151, 205, 216,
  220, 246n11
Ahmed Shah 43, 116, 203
aircraft, use of 217–18, 222–3, 246n18
Ali Khel 163, 176
Ali Masjid, fort 54, 102, 103, 113, 146,
  151, 157, 239n16
Amanullah, Emir 205–13, 219, 222–5
Amin, Hafizullah 230
Amritsar massacre (1919) 208, 245n7
Argyll, George Douglas Campbell,
  Duke of 5–6, 117, 136, 140, 233n6
Army of Retribution 99–105, 107–8,
  112, 114, 117, 119, 173, 226, 233n6
Army of the Indus, invasion of
  Afghanistan by xv, 31, 35, 37–42, 44,
  49, 51, 52, 53, 54, 57, 60, 90, 99, 100,
  107, 108, 146, 149, 155, 173

Auckland, George, Lord 8–10, 14,
  17–20, 25, 29, 31–2, 34–7, 39–40, 43,
  44, 48–50, 54, 57–60, 63, 72, 93–5,
  98, 113–14, 120, 157, 240n33
Avitabile, General Paolo di 101, 102
Ayub Khan 181, 187–96

Bagh 214–17
Baker, General Sir Thomas 167, 168,
  170, 172, 177, 178
Bala Hissar fort, Kabul 4, 50, 51, 66, 68,
  71, 74, 74, 76, 80, 94, 106, 110, 111,
  116, 155, 161, 164, 165, 171, 172–4,
  176, 177
Bala Hissar fortress, Peshawar 4
Baluchistan 39, 117, 196
Bamiyan 22, 49, 79, 105, 109, 110, 172
Barakzai family 20, 25, 32, 41, 44, 48,
  53, 110, 134, 135, 234n19
Barrett, General Sir Arthur 214
Biddulph, Major General Sir Michael
  150, 151, 155
Bokhara 18, 27, 105, 109, 182
Bolan Pass xvii, 38, 39, 40, 41, 54, 160,
  169
Bright, Major General Sir Robert 173,
  175, 177, 184, 242
Browne, Lt. Gen. Sir Samuel 149, 151,
  155, 156, 158, 159, 160, 167, 173,
  242n30
Brydon, Dr William 90, 91, 92
Burnes, Lt.–Col. Sir Alexander 18–25,
  32–3, 35, 38, 39, 49, 50, 53, 63, 65–9,
  71, 72, 78, 93, 94, 101
Burnes, Charles 67
Burnes, Dr James 33
Burrows, General George 186–193

251

cantonments, in Kabul 51–2, 62, 71, 72, 73, 75–6, 82
Carter, President Jimmy 230
Cavagnari, Sir Louis Napoleon 142, 146–7, 148, 157–61, 163–8, 173, 174, 182, 186
Chamberlain, General Sir Neville 146–7, 152, 155, 186, 242n24
Charasia 170, 184
Charikar garrison 69–70, 81, 112
Chelmsford, Lord 207, 211, 212, 217, 218, 219, 223
Chillianwala, Battle of (1849) 126
Churchill, Sir Winston 128, 201
Connoly, Captain John 163
Conolly, Captain Arthur 6–7
Cotton, General Sir Willoughby 37–41, 43, 46, 49, 57, 58, 59, 60
Cranbrook, Lord 145, 147, 148, 155, 156, 174, 184
Crocker, General George 215, 216, 218
Cumming, Lieutenant James Slater 101–2, 239n14

Dakka, bombing of 217, 218, 220
Dalhousie, James Ramsay, Marquess of 125, 126, 129, 132, 136
Daoud, Mohammed 229
Daoud Shah, General 175, 179
Dennie, Brigadier William 73, 89–90, 97
Disraeli, Benjamin 33, 139, 140, 141, 158, 160, 166, 182, 184
Dogra clan 45, 236n27
Dost Mohammed Khan, Emir 13, 17–21, 24–5, 29–33, 35, 37–8, 41, 42, 44, 46, 48–50, 53–4, 67, 68, 79, 86, 110, 113, 114, 115, 116, 129, 130, 132, 133, 144, 146, 149, 150, 155, 182, 203, 209, 235n19, 236n31, 240n33
Dulip Singh 120, 125
Durand, Major General Sir Henry Mortimer 31, 47, 48, 50, 200
Durand, Sir Mortimer 200
Durand Line xix, 159, 200–5, 208, 211, 220, 223, 224
Durranis 117, 127, 133, 234n19, 237n12
Dyer, General Reginald 208, 222, 224, 245n7

East India Company 6, 17, 63, 118, 128, 130, 235n1
  Board of Control 8, 29, 32, 93, 95, 131, 235n1
    Secret Committee 29, 37, 39, 235n1
  Court of Directors 119
  Sepoy Mutiny 128–9
Eden, Emily 8, 36–7, 45, 59, 120
Edwardes, Sir Herbert 130
Ellenborough, Edward, Lord 95–6, 98, 99, 105, 106, 109, 113, 114, 119
Elphinstone, Hon. Mountstuart 1–5, 13, 19, 31, 42, 126
Elphinstone, Major General William 2, 58–63, 65, 69–72, 74, 76, 80, 82, 83, 84, 86, 91, 93, 98, 105, 178, 180, 244n16
England, Major General Sir Richard 106

Fane, Lt. Gen. Sir Henry 35, 36, 37–8, 39
Feringhees 44, 48, 64, 78, 90, 92, 129, 150, 154, 164, 175, 202, 208, 236n22
Ferozepore, Punjab 35, 37, 122, 123, 149
Forward Policy 124, 135, 137, 138, 139, 140, 141, 143, 199

Gandamak 87, 105, 158
Gandamak, Treaty of 159–60, 164, 187, 196, 203, 210
Gardner, Colonel Alexander 121
Gates of Somnath 113
Ghazis 47, 48, 81, 82, 84, 85, 86, 87, 108, 112, 236n28
Ghazni 45, 46, 49, 54, 79, 107, 108, 116, 146, 171, 176, 181, 183, 184, 189
  siege of (1839) 46–51
Ghilzais 64, 127, 171, 183, 204, 234n19, 237n12
Goderich, Frederick John Robinson, 1st Viscount 244n28
Gordon Highlanders 239n16
Gough, Brigadier General Sir Hugh 120, 121, 123, 124, 126, 178
Great Game xv, 6–7, 12, 18, 199, 225
Griffin, Sir Lepel 182, 183, 186, 203

INDEX 253

Habibullah, Emir 205, 206–7, 222, 223
Haines, Sir Francis 166
Haines, General Sir Frederick 154, 189, 192
Halley, Captain Robert 'Jock' 222
Hardinge, Viscount Henry 120, 122–5
Hartington, Marquis of 184, 185
Haughton, Colonel John 70
Havelock, Major General Sir Henry 43, 97, 98
Herat 10, 11, 12, 14, 38, 39, 116, 133, 143, 155, 187, 199
  fall of (1863) 133
  siege of (1837–8) xvii, 11–19, 25, 27, 29, 31, 37, 39, 157, 199
Heytesbury, Lord 8
Hill, Major General James 174
Hobhouse, Sir John Cam 29, 32, 33, 58, 59, 62, 91, 95
Hughes, Brigadier General Francis 'Ted' 220
Humboldt, Alexander von 21
Hyderabad 40, 80, 118, 240–1n8

India
  widespread civil unrest xviii
  bureaucracy xx–xxi
  Persia as an invasion route to 7
  retiring native regiments go to (1841) 78
  hostilities with the tribes 127–8
  a Crown colony 131
  British withdrawal (1947) 127
India Office, London 131
Indian Mutiny (1857) 80, 99, 128–9, 131–2, 160, 169, 235n1, 242n30
Indus River 4, 17, 31, 39, 40, 62, 117, 144, 188, 202
Iraq, invasion of (2003) 34
ISAF (International Security Assistance Force) xiii, xiv

Jalalabad ix, 25, 52, 54, 65, 69, 73, 74, 82, 86, 89, 90, 92, 96, 98, 100, 104, 105, 113, 122, 148, 151, 155, 156, 157, 158, 177, 212, 219, 220, 225
Jamrud 129, 146, 147, 158
Jindan Kaur, Maharani 124–5
Jugdulluk Pass 86–7, 91, 108

Kabul
  Russia sends a military-diplomatic mission xvii–xviii
  Shuja ousted 5
  Pottinger appointed Envoy 13
  epicentre of Great Game intrigue (from 1837) 18
  Burnes abandons (1838) 24–5, 32
  British take control xiv, 49–53, 54
  British Resident murdered xv, xviii, 66–70
  siege 67–81
  evacuation treaty 82
  Nott and Pollock meet 108
  freed prisoners 109–10
  British retribution in 111–12
  Russian mission in 148
  British mission 157
  Cavagnari murdered xv, 165, 173, 182, 186, 226
  British occupy 172, 183
  battle-hardened garrison 194
Kabul Field Force 166, 170, 172, 174, 184
Kabul River 53, 97, 178
Kalanaki 228
Kalat-i-Ghilzai garrison 106, 155
Kamran, Shah 10, 14, 16
Kandahar 17, 25, 30, 34, 38, 41, 42–4, 46, 49, 54, 60, 73, 76, 77, 78, 82, 106, 133, 142, 147, 158, 166, 169, 181, 184, 186–7, 193
Karmal, Babrak 230
Kaufmann, General Konstantin 141, 145, 156
Keane, Lt. Gen. Sir John 38–42, 44–9, 54, 55
Khan, Abdul Rashid 46
Khan, Abdullah 68, 156
Khan, Afzal 134
Khan, Akbar 18, 67, 74, 79, 80, 81, 85, 86, 87, 91, 94, 96–8, 103, 104, 105, 108, 109, 115, 150, 180, 239n26, 240n2
Khan, Amanullah 66
Khan, Atta Mohammed 5, 77–8
Khan, Azam 134
Khan, General Faiz Mohammed 147
Khan, Fateh 116
Khan, Mohammed Anwar 215

Khan, Mohammed Nadir 228, 229
Khan, Nadir 206, 220, 221–2
Khan, Major Nakshband 164
Khan, Emir Sher Ali 115
Khan, Zaman 94
Kharak Singh, Maharajah 37, 45, 120
Khoord Kabul Pass 83, 84, 85, 111, 175
Khyber Pass xv, xviii, 13, 34, 54, 98, 100, 102–5, 113, 126, 129, 146, 148, 149, 152, 155, 157, 158, 176, 186, 201, 210, 215
Khyber Rifles 149, 207, 209, 211, 212, 214, 215, 216
Koh-i-Noor diamond 4, 5
Kohun-dil-Khan 42
Konegaon, Battle of (1818) 6
Kuzzilbashis 44, 236n26

Lahore 5, 34, 37, 45, 120, 124, 126
Lal Singh 122, 124, 125
Landi Kotal garrison 215, 216
Lawrence, Lt. General Sir George 80, 86, 105, 109–10
Lawrence, Sir Henry 13, 125–6, 129
Lawrence, John, Lord 129, 134, 135
Littler, General Sir John 122
Ludhiana 5, 19, 23, 30, 62, 95
Lumsden, Lieutenant Harry 238n28
Lytton, Lord Edward 139–49, 152, 155, 156, 158, 166, 167, 173, 174, 175, 181, 182, 184, 185, 186, 242n12

McCaskill, General Sir John 112, 113, 123, 240n32
Mackenzie, Brigadier General Colin 80, 86, 110
Maclaren, Colonel James 78, 79
Macnaghten, Lady Frances 52, 63
Macnaghten, Sir William Hay 9, 18, 19, 20, 21, 24, 25, 29, 30, 31, 33, 34–5, 37, 39–43, 45–54, 57–8, 60–1, 62, 63–71, 73–81, 86, 93, 95, 106, 111, 114, 135, 175
McNeill, Sir John 12, 14, 18, 26, 27
Macpherson, Sir Herbert 167, 170, 171, 175, 177, 178, 184
Mahsuds 127, 221
Maiwand, Battle of 187, 189–93, 195
Malakand garrison 201–2
Maratha Confederacy 6

Marathas 6, 233n7
Masson, Charles (James Lewis) 22–5, 101
Massy, Brigadier General William 178, 179, 180
Melbourne, William Lamb, Lord 62
Minto, Gilbert Elliot, Lord 2, 4–5, 6, 8
Mohammed Jan 176, 177, 178, 180
Mohammed, General Saleh 109, 206, 210, 219, 220
Mohammed, Shah of Persia 10–11, 12, 15, 16, 21, 26, 27, 29, 42
Mohammed, Vizier Yah 14, 15, 16
Mohmands 159, 218
Montagu, Edwin Samuel 211, 217, 224
Monteith, Lt. Col. Thomas 65, 97
Mudki, Battle of 123–4, 240n32
Musa Khan, Emir 176
Mushk-i-Alam, Mullah 176, 179

Napier, Major General Sir Charles 38, 52, 114, 117, 118, 119
Napoleon Bonaparte 2, 3, 7, 120, 142
Nasrullah (Habibullah's brother) 206, 208
Nesselrode, Count Karl Robert 26, 27
Nicholas I, Tsar of Russia 8, 11
Nicolls, Lt. General Sir Jasper 38, 59, 98, 99
North-West Frontier 4, 13, 23, 127, 129, 135, 145, 148, 159, 167, 202, 207, 212, 214, 246n18
Northern Alliance 228, 240n27
Nott, Major General William 60, 63, 64, 73, 76–9, 106–9, 114, 115, 116, 155, 173, 193, 239
Nowshera, Battle of (1822) 241n16

Outram, Lt. General Sir James 47, 132
Oxus River xviii, 138, 158, 200

Pacifico, Don 26
Palmerston, Henry John Temple, Lord 8, 10, 11, 12, 16, 17, 26–7, 29, 33, 95, 134
Panjdeh affair 200
Pashtun tribes xiv, xix, 1, 4, 42, 54, 94, 127, 128, 129, 130, 150, 151, 152, 154, 158, 159, 160, 204, 208, 210,

211–12, 217, 227, 228, 234*n*19, 236*n*31, 243*n*37
Pathan Revolt (1897) 201, 202
Paul I, Tsar 3
Peel, Sir Robert 8, 62, 93, 95, 99, 119, 239*n*20
Peiwar Kotal Pass 152, 153, 154
Peshawar 1, 3, 17, 20, 23, 24, 35, 45, 61, 78, 100, 101, 126, 129–30, 146, 210, 219
   capture of (1822) 241*n*16
Pollock, Major General Sir George 98–105, 108, 109, 111, 113, 114, 125, 173, 239*n*11
Pottinger, Major Eldred 13–16, 22, 25, 39, 70, 81, 82, 86, 91, 109
Primrose, Major General James 188, 192, 193, 194
Punjab 5, 6, 17, 34, 35, 100, 119, 121, 130, 136, 150, 208, 212
   annexation to the Indian Empire 20, 118, 125, 126, 236*n*27

Quetta 41, 54, 61, 106, 147, 150, 151, 194, 220
   British occupation 144

Raglan, Lord (*see* Somerset, Fitzroy)
Rawalpindi, Treaty of (1919, amended 1921) xviii, 223–6
Rawlinson, Sir Henry 78, 113, 115
Richards, General Sir David xiii–xv
Ripon, Lord 184, 185, 186, 192, 194
Roberts, Brigadier Abraham 57
Roberts, Lt. Gen. Sir Frederick ('Little Bobs') 57, 151–5, 160–1, 163, 166–81, 184–6, 193–7, 213
Roberts, Lady Nora 195
Roos-Keppel, General Sir George 207, 209–12, 219, 246*n*11
Royal Air Force (RAF) xv, xviii, 217, 219, 220, 222, 223
Royal Navy 8, 27
Russia
   expansionism xv, xvii, 6, 19, 95, 115, 124, 135, 137, 241*n*9
   military-diplomatic mission to Kabul xvii–xviii
   Panjdegh affair 200
   see also Soviet Union

Saddozais 24, 30, 34, 41, 43, 67, 94
Saidullah (the 'Mad Mullah') 201, 202
Sale, Lady Florentia 52, 72–3, 81, 82, 85, 86, 89, 91, 110
Sale, Major General Sir Robert 47, 48, 52, 54, 65, 71, 74–5, 76, 78, 86, 91, 96, 97, 100, 103–5, 109, 110, 114, 123, 186
Salisbury, Lord 139, 141, 142, 143, 144, 145, 150
Scientific Frontier 160
Sepoy Mutiny (1857) *see* Indian Mutiny
Shah, Mahmud 5, 116, 234*n*19
Shah, Nadir 224
Shah, Zahir 229
Shah Ali Raza, Colonel 205
Shah Shuja ul Mulk, Emir 3–5, 19, 23–4, 29, 30, 31, 34–5, 38, 40–4, 47, 49–51, 54, 57, 62, 66, 75, 79, 80, 83, 94–5, 96, 109, 114, 234*n*19, 235*n*5
Shakespear, Sir Richmond Campbell 109, 110
Shelton, Brigadier John 60–2, 66, 71, 72, 77, 91, 110, 179
Sher Ali, Emir 133–8, 140, 142–6, 148–9, 152, 154, 155–7, 174, 182
Sher Ali (Wali of Kandahar) 187
Sher Singh 45
Sherpa cantonment 172, 174, 175, 176, 178–81, 186, 213
Shia Muslims 10, 15, 17, 234*n*37, 243*n*37
Shutargardan Pass 167–8, 176
Sikh Army (Khalsa) 120–1, 122, 124, 125, 126
Sikhs 4, 6, 17, 20, 22, 34, 35, 45, 54, 100, 118–19, 120, 123, 124, 125, 177, 216
   national revolt (1848) 125–6
Simla 9, 25, 140, 163, 166, 185
Simla Manifesto (1838) 31–4, 52, 113, 157
Simla Proclamation 113, 240*n*33
Simonich, Count Ivan 11, 21, 27
Sind 24, 38, 40, 61, 117–18, 119, 136, 240–1*n*8
Singh, Nau Nihal 45
Singh, Ranjit Maharajah 4, 5, 17, 19, 22, 30–7, 45, 54, 101, 118, 119, 120, 121, 235*n*6
Skeen, General Sir Andrew 218

Somerset, Fitzroy (later Lord Raglan) 59
Soviet Union
  aid to Afghanistan xviii, 229
  invasion of Afghanistan (1979) xviii, 228
  withdrawal from Afghanistan (1989) 231
  see also Russia
Spin Baldak fort 220
Stewart, General 150, 154, 155, 166–7, 169, 173, 181, 186, 188, 194
Stoddart, Lt. Colonel Charles 7, 27
Sunni Muslims 15, 17, 234n16, 243n37
Sutlej River 6, 35, 39, 114, 121, 124

Tajiks 228
Taliban xiii, xiv, xv, 22, 42, 203, 228, 231
Taraki, Mohammed 229, 230
Tej Singh 123, 124
Tilsit, Treaty of (1807) 3
Timur, Prince 54, 60, 116
Trevor, Captain Robert 81
Turis 153, 243n37

United States
  diplomatic recognition of Afghanistan xviii, 246n28
  breaks Soviet air supremacy in Afghanistan 231
  invasion of Afghanistan (2001) 231
  September 2001 attacks 231

Victoria, Queen 30, 50, 118, 139, 150, 154, 192, 197, 201, 239n20
Vikevitch, Captain Jan 21–22, 23, 27, 234n29

Wade, Lt. Colonel Sir Claude 18, 20, 22, 23, 54, 95
Wapshare, Lt. General Richard 214, 220
Warburton, Colonel Sir Robert 148, 149, 150, 212
Waterloo, Battle of (1815) 2, 58, 114
Wazirs of Tochi 221
Wellesley, Arthur, Duke of Wellington 2, 31, 58, 59, 99, 117, 123, 237n6
Wild, Brigadier Charles Frederick 100, 102, 103
William IV, King 18

Yakub Khan 156–60, 168–9, 171–2, 176, 180, 184, 186, 245n35

Zaman Shah 116
Zia-ul-Haq, General Mohammed xxi